LEGAL BASIS FOR A NATIONAL SPACE LEGISLATION

SPACE REGULATIONS LIBRARY

VOLUME 3

The titles published in this series are listed at the end of this volume.

LEGAL BASIS FOR A NATIONAL SPACE LEGISLATION

by

JULIAN HERMIDA
Aerospace Consultant,
Halifax, NS, Canada

KLUWER ACADEMIC PUBLISHERS
DORDRECHT / BOSTON / LONDON

A C.I.P. Catalogue record for this book is available from the Library of Congress.

ISBN 1-4020-1913-0

Published by Kluwer Academic Publishers,
P.O. Box 17, 3300 AA Dordrecht, The Netherlands.

Sold and distributed in North, Central and South America
by Kluwer Academic Publishers,
101 Philip Drive, Norwell, MA 02061, U.S.A.

In all other countries, sold and distributed
by Kluwer Academic Publishers,
P.O. Box 322, 3300 AH Dordrecht, The Netherlands.

Printed on acid-free paper

Printed in the Netherlands.

To Florencia, Milton and Lucas

TABLE OF CONTENTS

INTRODUCTION

A. GENERAL BACKGROUND

> "The foremost goal of the international community in the area [of private space launch services] should be **to induce states to implement effective licensing procedures applicable to commercial ventures** for which state responsibility may exist.[1]"

1. PRIVATE SECTOR PARTICIPATION IN THE SPACE INDUSTRY

In the first decades of the space age, military and state security motivations indicated the direction of national space programs. Now the development of space activities depends essentially upon the possibility of recovering investments.[2] Private sector-driven commercial endeavors in outer space have been increasing exponentially and have experienced a significant quantitative growth over the last years. Spacefarers promote commercial participation of private companies in operations related to outer space, and, thus, the private sector is now increasingly providing satellite telecommunications, remote sensing, global positioning and space launch services directly to its customers.

In this context, overall revenues for the worldwide space industry amounted to US$ 82 billion in 2001[3]. In the late 1990's the transponder demand, in particular Ku- band transponders, was consistently on the rise due to the escalated utilization of geostationary satellite transponders[4]. Global positioning systems have been playing an increasingly important role in navigation, and remote sensing systems are mapping and documenting nearly

---

[1] E. A. Frankle & E. J. Steptoe, "Legal Considerations Affecting Commercial Space Launches From International Territory", (1999) 50 *IISL* at 10. Emphasis added.

[2]H. L. Van Traa-Engelman, *Commercial Utilization of Outer Space* (Dordecht: Martinus Nijhoff Publishers, 1993) at 18 [hereinafter "Commercial Utilization of Outer Space"].

[3] The 2002 State of the Space Industry, International Space Business Council, February 2002, http://www.spacebusiness.com/ accessed July 18, 2002.

[4]Satnews, February 28, 2002 http://www.satnews.com/stories2/1mar2002-2.html accessed July 18, 2002.

the entire world to the centimeter level[5]. The space launch market also experienced a steady and constant growth during the 1990's[6]. Although there was a slowdown in the satellite and space launch markets at the beginning of the new millennium with some programs sent to bankruptcy in the United States, and post-September 11 security measures, when the global economy rebounds, demand for next-generation satellite applications and launch services is expected to grow again at pre-September 11 projections[7].

This increasing shift to the private sector has provoked a complete realignment of the focus of regulation toward the adoption of domestic laws aimed at fostering and governing the realization of private sector activities in outer space. The enactment of national space legislation geared toward regulating the space activities of private entities is an imperious necessity since pre-commercial space age regulation has proved inadequate to deal with the complex and sophisticated endeavors of private firms in outer space[8]. A small number of spacefaring nations have been implementing a comprehensive domestic regulatory framework to govern their national activities in outer space[9] and are now urging[10] those countries with developing space capabilities, in particular those that are currently considering the implementation of space transportation systems[11], to adopt national space legislation.

With the view to contextualizing the present study, consideration must first be given to the different paradigms of space development throughout its history and the parallel evolution of Space Law. Although any attempt to divide history into distinct segments is not a precise enterprise, this

---

[5] Global Positioning Satellite Market Expanding Rapidly, Space Daily, September 29, 1997, http://www.spacedaily.com/news/gps-97c.html accessed December 17, 2001.

[6] Satellite Industry Association/Futron Satellite Industry Indicators Survey 2000/2001, June 2001 http://www.sia.org/satelliteinfo.htm accessed July 18, 2002.

[7] Global Demand for Satellite Systems in the Commercial, Military, and Science Sectors. Northern Sky Research http://www.satnews.com/frames.html accessed October 10, 2003.

[8] J. Hermida, *Commercial Space Law: International, National and Contractual Aspects* (Buenos Aires: Ediciones Depalma, 1997) at 17 [hereinafter "Commercial Space"].

[9] F.G. von der Dunk, Private Enterprise and Public Interest in the European 'Spacescape' Towards Harmonized National Space Legislation for Private Space Activities in Europe (Leiden, IIASL, 1999) [hereinafter "Spacescape"].

[10] I. Volk & A. Bauzá Araujo, "Space Transportation", in M. Folchi, ed., 40 Años de ALADA (Buenos Aires: ALADA, 2000) at 103.

[11] These include Brazil, Japan, Israel and Argentina, among others. E. A. Frankle & E. J. Steptoe, *supra* note 1 at 4.

examination is useful since it will permit us to identify the major features of each paradigm and to define the theoretical approach to conduct the instant study.

2. THE OUTER SPACE EXPLORATION ERA

The first phase of the space age, whose starting point was the launch of Sputnik 1 in 1957, is characterized by exploration activities carried out by the then superpowers in search for prestige and dominance on Earth[12]. The space programs of the United States and the Soviet Union were based on the premise that the exploration of outer space represented a competition whose main reward was political prestige on Earth.[13]

From a commercial standpoint, this philosophy implied that States acted under the parameters of a war economy, with generous budgets and without taking into account the financial aspects of their activities and the relationship between costs and benefits[14]. Furthermore, under this conception both in the United States and the Soviet Union criticism to the space programs was viewed as country treason, instead as simple requests to rectify those activities toward objectives of more scientific value or to economic rationality. Thus, both US and Soviet space programs emphasized quantity over quality and practically every proposal of relative scientific merit was soon accepted. During this stage there were no endeavors of a commercial nature[15].

During this period Space Law centered around topics and issues of international law[16]. Strongly influenced by the political context of the cold war, International Space Law -created through the search for the minimum consensus between the then world superpowers- concentrated mainly on the regulation of the exploration of outer space for peaceful purposes[17]. Thus,

[12] J. Hermida, *Norms governing launch services by NASA and commercial US private companies*, (LL.D. Thesis, Catholic University of Cordoba, Doctorate of Laws Thesis 2000) [unpublished] [hereinafter "Launch Services"] at 6.
[13]M. Couston & L. Pilandon, *L'Europe Puissance Spatiale*, (Bruxelles: Bruylant, 1991) at 92.
[14] E. Diamond, *The Rise and Fall of the Space Age* (New York: Doubleday & Company, Inc., 1964) at ix.
[15] *Ibid.* at ix.
[16] M. A. Ferrer (h), *Espacio aéreo y espacio superior* (Córdoba: Dirección General de Publicaciones, 1971) at 396.
[17] I. Vlasic, "A Survey of the Space Law Treaties and Principles Developed through the United Nations", (1995) 38 *IISL* at 324.

military and humanitarian issues became the almost exclusive concerns of this field[18].

The United Nations played an essential role in the development of Space Law during this first stage. In 1958 the UN General Assembly created the Committee on the Peaceful Uses of Outer Space (COPUOS), where International Space Law would be discussed and codified[19]. COPUOS was divided into two subcommittees: the Legal Subcommittee and the Scientific and Technical Subcommittee. However, since COPUOS consisted mostly of members from capitalist countries, only after a few years of lengthy negotiations between the Soviet Union and the United States that led to an increase of socialist states from 18 to 28 did COPUOS actually start to function[20].

The United States held that COPUOS decisions should be taken by a simple majority and then they should be sent to the General Assembly for approval. The Soviet Union was against this procedure, since the United States and its allies outnumbered the socialist states.[21] Thus, COPUOS adopted the consensus procedure for decision making, especially thanks to the efforts of Manfred Lachs, who at that time presided over the Legal Subcommittee. Consensus in COPUOS is conceived as the search for the common ground in a debate by means of a scientific discussion of the problem until an agreement is reached. Consensus means the acceptance of the discussed option in all its scopes, which implies a common feeling by those that choose it.[22]

Through this procedure, COPUOS produced the five international space treaties[23]: (i) Treaty on Principles Governing the Activities of States in

---

[18]The first works on Outer Space Law date from the beginning of this century. These first analyses belong to Emile Laude (1910) and Vladimír Mandl (1932). The most complete and consistent works appeared in the 1950s. Among others, the following must be highlighted: Andrew G. Haley, Eugène Pépin, Isabella H. Ph. Diederiks-Verschoor (then de Rode-Verschoor), and Aldo Armando Cocca. See "Commercial Space" *supra* note 7 at 13.

[19] N. M. Matte, *Space Policy and Programmes Today and Tomorrow* (Montreal: McGill University, 1980) [hereinafter "Space Policy and Programmes"]at 21.

[20] *Ibid.* at 21.

[21]*Ibid. 21.*

[22]A. A. Cocca, *Desarrollo progresivo del Derecho Internacional* (Buenos Aires: Consejo de Estudios Internacionales Avanzados, 1991) at 47.

[23] The consensus procedure was abandoned when COPUOS dealt with the declaration on the "Principles Governing the Use by States of Artificial Earth Satellites for International Direct Broadcasting Television." Then COPUOS returned to the

the Exploration and Use of Outer Space, Including the Moon and Other Celestial Bodies (1967); (ii) Agreement on the Rescue of Astronauts, the Return of Astronauts and the Return of Objects Launched into Outer Space (1968); (iii) Convention on International Liability for Damage Caused by Space Objects (1972); (iv) Convention on Registration of Objects Launched into Outer Space (1975); and (v) Agreement Governing the Activities of States on the Moon and Other Celestial Bodies (1979)[24].

In the first two decades of work devoted to the creation of Space Law, the United Nations achieved important results, providing a fairly general framework for the activities in outer space. Except for the Moon Agreement, all space law treaties have been widely ratified by the international community[25].

---

consensus procedure for the adoption of the three following declarations: "Principles Relating to Remote Sensing of the Earth from Space" (1986), "Principles Relevant to the Use of Nuclear Power in Outer Space" (1992), and "Declaration on International Cooperation in the Exploration and Use of Outer Space for the Benefit and in the Interests of All States, Taking into Account the Needs of Developing Countries" (1996).

[24]These are: (i) Treaty on Principles Governing the Activities of States in the Exploration and Use of Outer Space, Including the Moon and Other Celestial Bodies, 27 January 1967, 610 U.N.T.S. 205, 18 U.S.T. 2410, T.I.A.S. No. 6347, 6 I.L.M. 386 [hereinafter the "Outer Space Treaty"]; (ii) Agreement on the Rescue of Astronauts, the Return of Astronauts and the Return of Objects Launched into Outer Space, 22 April 1968, 672 U.N.T.S. 119, 19 U.S.T. 7570, T.I.A.S. No. 6599, 7 I. L.M. 151 [hereinafter the "Rescue and Return Agreement"]; (iii) Convention on the International Liability for Damage Caused by Space Objects, 29 March, 1972, 961 U.N.T.S. 187, 24 U.ST. 2389, T.I.A.S. No. 7762 [hereinafter the "Liability Convention"]; (iv) Convention on Registration of Objects Launched into Outer Space, 14 January 1975, 1023 U.N.T.S. 15, 28 U.S.T. 695, T.I.A.S. No. 8480 [hereinafter the "Registration Convention"] and (v) Agreement Governing the Activities of States on the Moon and Other Celestial Bodies (1979).

[25] As of 1 January 2003, the Outer Space Treaty has been ratified by 98 States and signed by 27 others; the Rescue and Return Agreement has been ratified by 88 States and signed by 25 others. One international intergovernmental organization has declared its acceptance. The Liability Convention has been ratified by 82 States and signed by 25 others. Two international intergovernmental organizations have declared their acceptance. The Registration Convention has been ratified by 44 States and signed by 4 others and two international intergovernmental organizations have declared their acceptance. The Moon Agreement has only been ratified by 10 States and signed by 5 others. http://www.oosa.unvienna.org/FAQ/splawfaq.htm#index accessed May 20, 2003.

3. COMMERCIAL INNOCENCE AGE

This phase, which began in 1962 with the creation of COMSAT[26], followed by the launch of INTELSAT's Early Bird -the first commercial purpose satellite- in 1964 is characterized by an almost *naïf* conception of commercial space activities. Under the prevailing political philosophy, it was believed that the exploitation of outer space offered infinite commercial opportunities, which assured astronomical returns[27].

The main protagonists of this phase were the States and the international intergovernmental organizations. The participation of private sector companies was relegated to a secondary role. Private companies acted as State contractors and carried out almost exclusively Earth activities. It was the States that assumed the business risk of all commercial space endeavors. During this stage, the major intergovernmental telecommunications organizations, such as INTELSAT[28], INMARSAT[29] and INTERSPUTNIK[30] were created, as well as the European Space Agency[31].

---

[26] Comsat Corporation was created by the Communications Satellite Act of 1962 and incorporated as a publicly trade company in 1963. K. Katkin, "Cable Open Access and Direct Access to Intelsat", (2002) 53 *Case W. Res.* at 87.

[27] "Space Policy and Programmes", *supra* note 18 at 115.

[28] INTELSAT was created in 1964 on an interim basis by eleven nations who joined together to establish a global commercial communications satellite system that would be available to all nations for expanded telecommunications services on a non-discriminatory basis. In February 1973 the final structure of INTELSAT became effective. Recently it underwent a major privatization transformation. D. Wear, "INTELSAT: Evolving to Meet the Challenges of a New International Telecommunications Marketplace", (1995) 38 *IISL* at 131.

[29] In 1999 INMARSAT was privatized as of 1999. The international governmental organization continues –in a residual fashion- and is now referred to as IMSO.

[30] The Agreement which established INTERSPUTNIK was signed on November 15, 1971, and entered into force on July 12, 1972. INTERSPUTNIK was founded as an international intergovernmental organization, consisting of former Soviet states. Originally formed by the COMECON countries, INTERSPUTNIK changed its composition after the fall of the Soviet Union. On June 2, 1997 Lockheed Martin's Space & Strategic Missiles Sector and the INTERSPUTNIK International Organization of Space Communications formed a joint venture company to provide worldwide communications services. M. Hoskova, "Intersputnik - New Legal Developments", (1995) 38 *IISL* at 139.

[31] In 1972 during the European Space Conference, European states, led by France, agreed to create only one European Space Agency and dissolve the two existing ones--ELDO and ESRO--which up to then had proved incapable of offering a coherent space policy.

The economic analyses made during this phase predicted unlimited returns in the exploitation of outer space[32]. The underlying idea was that space technology would evolve so as to permit an infinite use of space resources. Under this political philosophy, once the technical problems were overcome, the exploitation of outer space would become free of any obstacles, which would give rise to unlimited economic opportunities to the benefit of all mankind[33].

This phase witnessed the rupture of the exclusive monopoly of the United States and the Soviet Union[34] and the emergence of other national or regional space programs, such as the European one under the leadership of France[35], and the Chinese one, among others[36].

During this stage, the focus of Space Law shifted from international law to the institutional aspects of the main intergovernmental organizations and to the domestic law of the United States[37]. During this stage, known as Organizational Space Law[38], specialized authors devoted mainly, among other aspects, to the analysis of the legal framework of intergovernmental institutions[39] as well as to United States domestic law[40]. During this period, United States domestic rules basically referred to authorizations to carry out space activities, liability issues and the use of space facilities, among other matters[41]. The law of intergovernmental organizations dealt mainly with the

---

[32] "Space Policy and Programmes", *supra* note 18 at 115.

[33] C. B. Christensen, "Economics and Regulation of Space Activities" in J. S. Greenberg & H. R. Hertzfeld, Henry R (eds.), *Space Economics* (Washington, DC: AIAA, 1992) at 215.

[34] A. S. Piradov, "L'Union Soviétique à l'origine de l'élaboration du droit de l'espace et de la création du Comité des utilisations pacifiques de l'espace extraatmosphérique" in A. S. Piradov (ed.), Droit Spatial International (Moscou: Progress Publishers, 1976) at 97.

[35] L. Laidet, "The French Space Program", in W. C. Thompson & S. W. Guerrier, eds., *Space: National Programs and International Cooperation* (Boulder: Westview Press,) at 63

[36] "Commercial Space", *supra* note 7 at 18.

[37] Kayser, *Legal Aspects of Private Launch Services in the United States*, (LL.M., Thesis, McGill University, 1991) [unpublished], at 136 [hereinafter "Private Launch"].

[38] "Commercial Space", *supra* note 7 at 16.

[39] "Launch Services", *supra* note 11 at 14.

[40]Goldman Nathan C., *American Space Law: International and Domestic* (Ames: Iowa State University Press, 1988).

[41] Laws and regulations dealing with satellite telecommunications services were nonetheless quite developed. The approach followed by the United States in this field

rights and obligations of the members of the organizations and with the regulation of the relationship between the organization and other entities[42]. During this stage neither the norms of international organizations nor the laws of the United States concentrated on the regulation of private entities[43]. Towards the end of this period, COPUOS' work decreased. It failed to adopt a new international convention on space law and its role as the main forum for discussion of space law norms also diminished.

4. COMMERCIAL SPACE AGE

The Commercial Space Age began in the 1980's and it is characterized by the search for economic profits through commercial space activities. The commercial space endeavors are no longer the exclusive prerogative of States and intergovernmental organizations. Private sector companies are now increasingly involved directly in commercial space activities[44]. During this phase, the Cold War ended and the role of States in the space race changed radically[45]. Private sector companies assumed a direct and more active role in the provision of goods and services directly to their customers. This process of maximization of profits from space resources introduced a policy of costs as a tool for decision taking in commercial space endeavors[46]. This phase also witnessed the incorporation of many new countries, including some developing States, to the group of nations with space capabilities[47].

---

was to declare the Communications Act of 1934 applicable to space telecommunications. After this declaration made by the FCC in 1970 many specific satellite telecommunications regulations were adopted.

[42] "Launch Services", *supra* note 11 at 6

[43] The main exception is the adoption of authorization rules for the launch of satellites. Goldman, *supra* note 39 at 1.

[44]N. Jasentuliyana, "Opening Remarks", (1995) 38 *IISL* at 317.

[45] However, States continue to play a significant role, but private sector companies have an active participation in the provision of goods and services directly to their costumers. Additionally, spacefaring nations started the reduction of military budgets and transferred the costs of space research and development to the private sector. S. Sanz Fernández de Córdoba, "Changing Basic Space Laws: Popularity, Pragmatism and Historical Lessons", (1993) 39 *IISL* at 329.

[46] G. A. Hazelrigg, "Cost Estimating for Technology Programs" in J. S. Greenberg, Joel S & H. R. Hertzfeld, Henry R. (eds.) *Space Economics* (Washington, DC, AIAA, 1992) at 97.

[47]M. L. Smith III, *International Regulation of Satellite Telecommunications after the Space WARC*, (D.C.L., Thesis, McGill University, 1989) [unpublished], at 70.

During this phase the works on Space Law focused on the regulation of commercial space endeavors from a business law perspective, stressing on the evolutionary, mixed and multidimensional aspects of commercial space activities[48]. As a response to the increasing commercial exploitation of outer space by US and European private entities, several States enacted specific domestic legislation to regulate the new space business ventures. These developments have widely attracted the attention of authors and policy makers around the world[49], and the enactment of domestic space legislation aimed at regulating space activities of private entities constitutes the latest stage in the evolution of Space Law[50].

3. THEORETICAL FRAMEWORK

The issues dealt with in this book will be analyzed through a myriad of theoretical angles, which includes Commercial Space Law, Comparative Law, and Law Reform and participatory theory.

For the examination of national space law in the spacefaring nations the present study will resort to the methodology, elements and techniques of Comparative Law[51]. The purpose of the comparative study is to illustrate the different responses to the international obligations assumed by States with respect to the regulation of national activities in outer space. The underlying premise behind this comparison is that there are common denominators in all the countries which implemented domestic space legislation and that these common denominators should constitute the basis for the legal framework to structure future national private space activities.

The value of employing the comparative law method lies not merely in the identification of common elements but in its facilitating a comprehensive understanding of the context and the factors influencing the

---

[48] M. Couston, *Droit Spatial Économique* (Paris: SIDES, 1994) at xxvii.

[49] "Spacescape", *supra* note 8 at 1.

[50] *Ibid.* at 1.

[51] We conceive Comparative Law as the comparative study of selected aspects of different countries, whether they belong to different legal systems or not. Authors who understand that comparison may only be carried out between legal aspects of different legal systems usually depart from the fiction that common law is a fixed and immutable body of law which has always existed and is substantively uniform throughout the common law world. P. De Cruz, *Comparative Law in a Changing World*, (London: Cavendish Publishing Limited, 1995) at 224 [hereinafter "Comparative Law in a Changing World"].

adopted domestic solutions[52]. Central to this approach is our conception that the licensing and authorization of space activities do not need new rules aiming at their harmonization as advocated by certain recent studies on comparative national space law[53]. What is actually needed is a wider tolerance and respect for differences between laws of different countries which reflect differences in the substantive concerns and political values. Also implied in this approach is our conception that the structure of the regulation of the distribution of risks and liability, in particular in space transportation, tends to converge in all jurisdictions[54] due to the consequences derived from international obligations assumed by States and by the consolidation of the launch services market[55].

The proposal for the foundations of a national space law for the regulation of commercial space activities will be analyzed in light of Law Reform and participatory theory. Law reform is conceived as a multifold dynamic process, which requires a national effort based on high level of State and private sector participation[56]. Law reform is the instrument for guiding and legitimizing the processes of change in society with due account of reconciling diverse interests[57]. Under this conception a law reform must

---

[52]P. De Cruz, *A Modern Approach to Comparative Law*, (Deventer: Kluwer 1993) at 34 [hereinafter "Modern Approach"].

[53] M.Gerhard & K.U. Schrogl, Report of the Working Group on National Space Legislation, Project 2001, May 2001 at 20.

[54] Convergence is the evolution of legal institutions of different legal systems in the same direction up to a point where the legal institution of one system resembles the other and the legal norms, principles, and scholarly comments of both are used in equal measure and even regarded as authoritative as each other. "Modern Approach", *supra* note 51 at 34.

[55] M. Bourély, "Space Law and the European Space Agency" in N. Jasentuliyana (ed.), *Space Law, Development and Scope* (New York: Praeger, 1992).

[56] Participatory law reform –also known as sociolegal law reform shifts the center of the current paradigm from the expert to the social actors. It rests on sociolegal research and employs sociological methodologies. It also implies a radical departure from a vision that presupposes that expertise backed by pseudo empirical evidence in centralized countries is the best vehicle for law reform. By placing its emphasis on the engagement of the public at large and by focusing on social relations and social process, participatory Law Reform accepts the idea that all of society's normative arrangements constitute legal normativity, worthy of consideration for meaningful and successful social change. R. Macdonald, "Recommissioning Law Reform 35" *Alberta L. Rev.* at 851; J. R. Nolon, "Fusing Economic and Environmental Policy: The Need for Framework Laws in the United States and Argentina" 13 Pace Envtl. L. Rev. 1996 at 726.

[57] I. F.I. Shihata, "The Role of Law in Business Development", (1997) 20 *Fordham Int'l L.J.* at 1578.

necessarily rest on three basic pillars: (i) adequate rules, (ii) appropriate processes through which those rules are made and enforced and (iii) well functioning public institutions appropriately staffed with trained individuals[58]. Throughout our discussion on the proposal and recommendations for the domestic legal scenario we will deal with all these three basic pillars and we will identify the major consequences of proposed changes to the existing scenario. Participatory theory requires that an act or any other regulation contemplate procedures allowing the industry, those affected by the law and the general public to participate in the elaboration of the regulations.

Outer space issues will be analyzed under Commercial Space Law concepts and categories of analysis. Commercial Space Law presents well defined characteristics and unique categories of analysis, which enable the examination of sophisticated systems and problems[59].

It has been suggested that there is a *lex mercatoria spatialis* formed by the uses and customs arising from commercial space practice, as well as the principal common legal solutions adopted in the main spacefaring nations[60]. The analysis of this *lex mercatoria spatialis* has given rise to the discipline of Commercial Space Law[61]. The *lex mercatoria spatialis* is a theoretical construction and it is not a binding or formal system of laws adopted by any lawmaking forum[62]. The underlying idea is that the practice

---

[58] *Ibid.* at 1583.

[59] "Commercial Space", *supra* note 7 at 19.

[60] K. H. Böckstiegel, "The Law Applicable to Contracts" (1982) 25 *IISL* at 207 [hereinafter "Contract Law"].

[61] Mireille Couston defines the nature of Commercial Space Law, which she prefers to call Economic Space Law, as evolutionary, mixed and multidimensional. The evolution of this field of law is strongly tied to the constant increase of commercial space activities and the participation of private sector entities, fostered by a progressive change in the role of states. The mixed aspect is given by (i) the diversity of the legal domain, where private and public law rules are entangled and (ii) by the diversity of the economic domain due to the fact that economic space relations, unlike governmental exploration activities, belong both to the Macroeconomics and Microeconomics order. The multidimensional aspect stems from the essence and origin of the rules which govern such economic space relations. M. Couston, *supra* note 47 at xxvii.

[62] Around two decades ago, Karl-Heinz Böckstiegel held that from an academic point of view this idea of the existence of a new lex mercatoria was extremely interesting, but from the practical point of view, it had not been accepted by any of court. "Contract Law", *supra* note 52 at 207. At present this situation has not changed. This, however, does not invalidate the conclusion that there is a certain consistency in the

derived from space activities has developed a typical and standardized system of rules. These rules encompass both the norms which are common to the majority of the domestic space law systems and those which have been developed by the different participants (private, public or semipublic companies, international intergovernmental organizations, etc.) engaged in rendering and acquiring commercial space services as reflected in the contracts and other types of agreements which they execute[63].

Authors and commentators have identified certain common features which, although not always present, may be said to characterize Commercial Space Law[64]. These features include: internationality[65]; tendency to the application of US Space Law solutions[66]; flexibility of contractual undertakings[67]; rigid risk allocation system[68]; and existence of different legal status of Commercial Space Law parties[69].

---

regulation of current commercial space activities, which allows its description under the denomination of *lex mercatoria spatialis*.

63 "Commercial Space", *supra* note 7 at 18.

64 P. Kahn, ed., *L'Exploitation commerciale de l'espace: droit positif, droit prospectif*, (Dijon: Litec Credimi, 1992) at 96.

65 Commercial Space Law is essentially international, since its object, i.e., commercial space activities, is international by nature. L. Peyrefitte, *Droit de l'espace*, (Paris: Précis Dalloz, 1993) at 18; P. Kahn, "Situations d'un droit commercial spatial", in P. Kahn, ed., *L'Exploitation commerciale de l'espace: droit positif, droit prospectif*, (Dijon: Litec Credimi, 1992) at 96.

66 For example, it is usual to find the rules developed by NASA for its launch agreements in the launch contracts used by its European competitor Arianespace. These national rules thus become part of Commercial Space Law. This phenomenon is due to two essential factors. First, most space agreements, especially the first ones, were executed by NASA or other participants from the United States, which provided the subsequent agreements with the characteristics of this country's legal system. Second, the United States has exercised a powerful influence on almost all commercial activities. Such influence is not foreign to the exploitation of outer space. P. L. Meredith & G. S. Robinson, *Space Law: A Case Study for the Practitioner* (Dordrecht, Martinus Nijhoff, 1992) at 7 [hereinafter "Case Study"].

67 Since every space activity entails high risks, and since the environment of outer space is hostile, the parties to a commercial space venture, in particular the ones that have a stronger bargaining power, tend to limit and soften their obligations, generally committing only to use their best efforts, instead of guaranteeing a certain result. In general, this loosening of the undertaken obligations is represented by the so called best efforts principle, which is sometimes defined in space transport agreements as "Diligently working in a good and workman-like manner, as a reasonable, prudent manufacturer of launch vehicles and provider of launch services." By means of the best efforts principle, the parties refrain from promising the accomplishment of their respective obligations, committing themselves only to making their best efforts to

B. SCOPE OF THIS WORK

1. PURPOSE

The private space industry necessitates an adequate legal framework for the regulation of its commercial activities. Many countries with an emergent private sector industry and without a specific domestic legal framework or with pre-commercial space age regulations are unable to provide a regulatory solution to adequately deal with the complex and sophisticated endeavors of private firms in outer space.

This study thus attempts to propose the fundamental regulatory policy basis for a future domestic legislation governing private and other non governmental space activities, in particular, space launch services, in those states which have not yet crafted their national space regulatory framework. The purpose is not to draft a legislative bill or to exhaustively define all the issues to be addressed in a domestic regulatory system. This is so because it is our understanding that law reform aimed at developing a comprehensive space legal framework at the national level must result from full participation of all space players and those likely to be affected by the reform. In this

---

achieve success. This principle is associated with both a reduction and a waiver of liability. B. Schmidt-Tedd, "Best Efforts Principle and Terms of Contract in Space Business" (1988) 31 *IISL* at 330; T. L. Masson Zwaan, "The Martin Marietta Case or How to Safeguard Private Commercial Space Activities" (1992) 35 *IISL* at 247 [hereinafter "Martin Marietta Case"].

[68] The distribution of risks in commercial space practice, especially in the area of space transportation, offers relatively consistent characteristics. This makes the risk allocation and assignment of liability a complex system with well-defined features and important financial consequences for the outer space industry. J. Hermida, *Legal Aspects of Space Risk Management. The allocation of risks and assignment of liability in commercial launch services*, (LL.M., Thesis, McGill University, 2000) [unpublished], at 4 [hereinafter "Space Risk Management"].

[69] Although at first space activities, even commercial ones, were almost exclusive monopolies of the states, at present, aside from the private sector, other participants, who are under different legal status have engaged in the exploitation of outer space. These include mixed entities, international intergovernmental organizations, every kind of governmental entities, national space agencies, international non-governmental institutions, joint ventures and national and international joint consortiums, among many others. The participation of these entities provides the agreements with particular characteristics arising from the nature of their organization. Thus, Public Law, International Law and domestic law matters -which reflects the different legal systems which the parties belong to- are incorporated in the contracts entered into by the protagonists of space activities.

respect, we follow Roderick Macdonald's rejection of expert law reform, which generally reduces the law reform process to the identification of problems cast in doctrinal terms and to the mere elaboration of a draft statute by experts and high-profile participants in the relevant field of law[70]. This traditional conception of law reform does nothing but reinforce a rational, non participatory and apolitical conception of the law reform process.[71] Thus, the objective of this book is to propose the main guidelines which should conform a general common legislative agenda for the discussion and elaboration of national space law in those States which currently lack a domestic space legal scenario.

We will also explore other avenues to deal with the problem generated by the emergence of a private space industry and the inadequacy of the domestic scenario -in those countries that have not adopted specific national rules- other than our proposed general common legislative agenda. Therefore, we will also delve into the proposals postulated in the legal literature and those advanced by international legal associations, such as the studies that advocate for the necessity to harmonize authorization, supervision and indemnification, and those which propound the adoption of new international rules or new interpretations of existing ones with the alleged purpose of adjusting the international legal framework to the needs of the private space industry.

Thus, as explained below in more detail, this book will first explore the international principles adopted in the *Corpus Juris Spatialis* to determine the nature, extension and content of the existing international foundations for the adoption of legislative or political measures at the national level. Second, it will focus on the experience of the countries which have adopted specific domestic legislation in the space field and on the alternative scenarios to our model of a common legislative agenda. Third, it will explore the main characteristics of the existing legal framework governing space activities in Argentina, which is analyzed as a case study since Argentina is one of the countries which has an important rising private sector space industry and its current domestic space laws are considered insufficient to adequately deal with the regulation of private space endeavors. Finally, it will propose the fundamental regulatory policy issues for a general common legislative agenda for those countries which have not yet adopted a –or have adopted an inadequate- domestic legal scenario.

---

[70] Macdonald, *supra* note 55 at 848.

[71] *Ibid.* at 848. Macdonald holds that: "this conception of presumes that the highest type of law, and the only type which is worthy of the efforts of experts, is law that is made explicitly by an official body such as the legislature or its delegates."

This study is based on the premises that the international legal framework governing space activities provides the fundamental basis for national space legislations and that the legislative experience of the countries which have adopted a domestic space legal scenario presents a useful model for delineating the principal basis of national legislation for those countries without specific national regulatory framework. Specifically, it is postulated that the common denominators of the countries with a significant history of involvement in the regulation of space activities should form part of any future domestic legal scenario.

2. RELEVANCE OF TOPIC

This work comprehensively analyzes national space launch legislation for a non spacefaring context, i.e., for a non US and a non European context, in light of Law Reform and Participatory theories. It is conceived as a future-oriented projection to advance regulatory policy formulation in the field of domestic space legislation.

This book explores the legal basis of a domestic scenario for private commercial space activities under a myriad of theoretical perspectives and a multi-approach methodology. It is expected that this book will not only be relevant for the academic arena, but also for the governmental and corporate worlds, for this study is expected to provide national policy makers with an alternative perspective to approach the regulation of space activities and to offer the space industry the foundations to construct a model for structuring its future commercial space practice in those countries which currently lack a national space regulatory framework. In particular, the conclusions and findings of this book are expected to determine the international basic foundations for the implementation of national space legislation. Second, they will attempt to offer the identification of common denominators of all other domestic legal norms and a critical assessment of the solutions adopted by other countries which are involved in space activities. Third, they will deconstruct the existing norms in Argentina and indicate the areas which are not in compliance with the international scenario and the solutions adopted by most other countries. Finally, the conclusions and findings of the present work are meant to provide the basis and the starting point for the debate on the adoption of national space legislation in those countries which have not yet adopted a domestic space regulatory scenario.

3. ORGANIZATION

The starting point of the analysis will be the examination of the international legal framework[72]. In particular, we will examine whether there is an obligation to enact domestic legislation arising from the authorization and continuous supervision principle contained in article VI of the Outer Space Treaty[73] and the fundamental foundations for national legislation which derive from the international legal framework[74].

We will then analyze the domestic legislation of States that are engaged in the exploration and exploitation of outer space. The aims are: (i) to describe the main characteristics of domestic Space Law, (ii) to compare the different approaches followed by the states, and (iii) to identify the common elements in all domestic systems. We will emphasize our analysis on the regulation of space transportation[75].

We will then proceed to the examination of the laws and regulations applicable to Argentine space activities and the policy objectives and guidelines contained in the National Space Plan. Consideration will also be made to the organization of space activities in Argentina and the nature, structure and functions of governmental agencies involved in the space arena. Our research will focus on whether or not the existing regulations create a

---

[72] According to the nature of the problems to be examined in the present work, the adopted research design is explanatory, comparative and descriptive with a reconstruction and interpretation emphasis. Due to the complexity of the problem, we follow a multi-method approach whereby we triangulate different elements and aspects of the research, i.e., we intend to explain the richness and complexity of the research object by resorting to more than one methodological standpoint. In this respect, we will triangulate diverse theoretical perspectives, with categories of analysis, as well as, interpretation techniques and methodologies. F. Carlino, *Tesis de Maestría*, (Buenos Aires: Flacso, 1997) at 44.

[73] Outer Space Treaty, article VI.

[74] The aim is not to merely explain but to discuss, explore and interpret the selected aspects of the research problem. This implies resorting to a myriad of interpretative methods of International Law and interrelating the principles enunciated in the *Corpus Juris Spatialis*.

[75] We will resort to the methods and techniques of Comparative Law. Specifically, the methodology to be employed throughout our Comparative Law examination is the micro-comparison method, i.e., the comparison of selected legal aspects of different countries, some of which share the same legal family while others belong to different legal systems. "Comparative Law in a Changing World", *supra* note 49 at 224.

legal environment conducive to the development of the local private space industry[76].

The final chapter will be devoted to the formulation of a proposal for the adoption of national space law in those countries which have not yet adopted a domestic regulatory framework for the regulation of their private commercial space activities. This will be achieved by systematizing all the conclusions and recommendations made throughout the book. These will include those aspects directly and indirectly arising from the international legal scenario and those aspects which derive from the national models adopted by other States. As mentioned above, our formulation of this regulatory policy will be done in light of legal reform and participatory theories. These recommendations will aim at setting the stage for a research and legislative reform agenda which is expected to be conducive to an effective reform of the domestic space legal scenario in those countries which currently lack a national legal scenario or which presently have only an inadequate one.

4. SEARCH FOR A COMMON LEGISLATIVE AGENDA

Our search for a common legislative agenda in those states which have not yet crafted their national space regulatory framework is geared by three distinctive sets of hypotheses. The first one deals with the International Space Law basis and foundations for the implementation of domestic space legislation. The second one regards the development of domestic space law - with particular reference to space launch legislation – in the most important spacefaring States. The third set of hypothesis is concerned with the actual implementation of national space legislation, particularly space launch legislation, in Argentina, a paradigmatic country for the analysis of the development of national space legislation.

The first hypotheses are basically centered around the responsibility and liability provisions, as well as the principle of State authorization and

[76] The third chapter of this book has been envisaged as a case study in terms of E. Greenwood. The case study represents the observation of the characteristics of an individual unit –Argentine national space law. The purpose of this observation is to probe deeply and to analyze intensively the diverse phenomena that constitute the multiple features of the selected individual unit with a view to assessing the current status of the research problem and establishing the main guidelines for the formulation of a proposed solution. E. Greenwood, *Methodology of social research* (Buenos Aires: Paidos, 1973) at 117; L. Cohen & L. Manion, *Research Methods in Education* (London: Routledge, 1994) at 107.

continuing supervision consecrated in the Outer Space Treaty. First, we postulate that the international responsibility and liability regimes impose significant burdens for the States, which dictates the necessity to administer and deal with them at the national level. Specifically, we will hold that States should structure safety laws or other measures to minimize the risks derived from the space endeavors of their nationals and at the same time, States should adopt a risk distribution system to reallocate these risks and liability according to their political objectives in the space arena. Additionally, we will attempt to prove that the principle of authorization and continuous supervision does not create an obligation to pass domestic legislation in every State and that States are free to implement the form of such authorization and supervision, which does not necessarily imply the enactment of domestic legislation.

As concerns the development of domestic Space Law, our first hypothesis is geared to showing that States have implemented the principle of State authorization and continuing supervision in different forms. Nonetheless, we hold that there are common denominators in all these national jurisdictions. Specifically with regard to domestic space launch legislation this will serve the present study to prove that any legal framework aimed at governing launch services must necessarily address the issue of the allocation of risks and assignment of liability. Additionally, we will also attempt to demonstrate that, even if there are certain important differences, which stem from the different political objectives of each of country, all these risk sharing regimes present general common features derived from the fact that they all have been modeled after NASA's. Finally, we will intend to prove that new international rules or new interpretations of existing ones are not needed, for they risk imposing new and unnecessary burdens to the space launch industry and its customers. Rather, it will be submitted that national legislation should be encouraged in those States which have not yet formulated a comprehensive legal framework for the regulation of commercial private sector space activities of their nationals or from their territories and that the common denominators of the national laws of spacefaring countries should serve as the basis for said national legal framework.

With respect to the implementation of national space legislation in Argentina as a paradigmatic case for the examination of domestic space legislation, our pivotal hypothesis is that existing Argentine space laws are insufficient to govern current and projected space activities, especially in the area of space transportation. Also, we postulate that Argentine Space Law norms are geared toward providing the local space authorities with enough powers to impede the development of space projects and endeavors which do not coincide with national space policy. Drawing on the findings of our analysis of the possible extent of the national regulations in light of International Space Law, we postulate that most of the restrictions adopted in the Argentine domestic system create a legal environment which is hostile to the existing private space industry and which hinders the development of private sector endeavors, particularly in the space transportation field. Further, we intend to demonstrate that the additional authorizations required for satellite telecommunications, which aim at protecting the dominant satellite service provider in detriment of other existing and potential satellite telecommunications players, also impose important obstacles for new firms to access the Argentine satellite telecommunications market.

The main purpose underlying this research is that a creative approach to legislative reform, especially along the lines of the domestic legislation of the main spacefaring nations,[77] in light of law reform and participatory theories will serve to provide an adequate framework for future space launch ventures in those countries which have not yet adopted a domestic space regulatory scenario.

---

[77] P. Nesgos, *National Law and Commercial Activities in Outer Space: Specific Aspects of United States Law Applicable to Space Industrialization with Particular Emphasis on the Commercialization of Launch Vehicle Transportation*, (DCL, Thesis, McGill University, 1983) [unpublished], at x [hereinafter "National Law and Commercial Activities in Outer Space"].

CHAPTER I

THE INTERNATIONAL LEGAL FRAMEWORK

Introduction

This chapter is devoted to the examination of the international aspects which directly influence the implementation of national space law. In particular, it will analyze the requirements that the *Corpus Juris Spatialis* imposes on States for the adoption of national measures and the general foundations for the adoption of domestic norms on account of the obligations assumed at the international level.

The chapter is divided into three substantive parts. The first one deals with international responsibility and liability of States. Since it has been extensively written about this topic, the present study will concentrate only on those aspects of international responsibility and liability which bear direct consequences for the adoption of national space legislation.

The second part of this chapter is devoted to the analysis of the international foundations for the implementation of measures at the national level. It begins with a discussion of whether there is an obligation under the *Corpus Juris Spatialis* to enact national space legislation. Then it analyzes the scope and extent of the authorization and supervision principle with the view to elucidating the foundations for the implementation of domestic space norms. Our analysis of the authorization and supervision principle is done in the light of the mosaic of international space law principles in order to facilitate its contextualization.

The remaining of the chapter delves into the mandatory system for the registration of space objects and the ensuing substantive and procedural obligations of States to implement at the national level.

A. INTERNATIONAL RESPONSIBILITY AND LIABILITY

The international responsibility and liability regimes of States for the activities of their nationals in outer space are of significant importance for the States because of the consequences which they entail both at the international and national levels[77]. On the international plane these regimes impose significant burdens for the States[78]. On the national level these burdens dictate the necessity to administer and deal with all their consequences in many ways. In this respect, for example, the whole system of authorization and continuing supervision[79] is the necessary and logical consequence of the fact that states are internationally responsible for the activities of their nationals[80] and internationally liable for damage caused by their space objects[81]. Therefore, we will now examine the international legal regime of responsibility and liability with the purpose of determining the aspects which all States should take into account in the implementation of a national regulatory scenario for space activities. Two hypotheses guide this discussion. First, since this regime imposes enormous burdens on the States for the activities of their nationals, States must adopt a system which allows them to protect themselves from the consequences of those activities. Second, States must also adopt an adequate risk management scheme to permit the fulfillment of their space policy objectives.

The discussion departs from the examination of the international responsibility and the international state liability regimes under general Public International Law and then it focuses on these regimes under the specific norms derived from the international space law treaties. Since it has been extensively written about this topic[82], the present study will focus solely on those aspects which bear direct consequences for the adoption of national space legislation. Thus, all other aspects will be left aside or reviewed very briefly.

1. RESPONSIBILITY IN INTERNATIONAL LAW

Historically in International Law responsibility has developed around injury of aliens and the whole current system –embedded in rules of customary international law- still reflects, to a large extent, its early roots[83]. Under this conception, it was considered that a state incurred international responsibility

---

[77] Outer Space Treaty, articles and VII.
[78] Outer Space Treaty, article VII.
[79] Outer Space Treaty, article VI.
[80] Outer Space Treaty, article VI.
[81] Outer Space Treaty, article VII.
[82] See M. Williams, *Report: Review of Space Law Treaties in View of Commercial Space Activities* (International Law Association: London, 2000) at 16 [hereinafter "ILA Report"].
[83] G. Schwarzenberger, *International Law as Applied by International Courts and Tribunals* (London: Stevens & Sons Ltd., 1957) at 564.

when it abused or failed to protect aliens. As synthesized by Hackworth, "an unredressed injury to an alien is considered to constitute an injury to the State, which triggers international responsibility"[84]. Thus, for example, in the Youmans claim, the General Claims Commission held that the Mexican government's failure to exercise due diligence to protect the victim, a US national engaged by a foreign corporation to work in Mexico, who was killed by Mexican laborers and soldiers, entailed Mexican international responsibility[85]. In the Corfu Channel Case, the International Court of Justice attributed international responsibility to Albania by presuming its knowledge of the existence of a peril and for its failure to advise the United Kingdom government of such danger[86].

In the second part of last century, international state responsibility centered around international rules regarding violation of a primary international obligation, which gave rise to a secondary obligation to make reparation for aliens' injuries[87]. The current notion of state responsibility is a "comprehensive regime of the law of obligations, covering general principles of States' international responsibility, including primary rules that establish all types of internationally wrongful acts attributable to a State and secondary rules that flow as a legal consequence from a State's breach of an international obligation, whatever its origin and regardless of whether aliens or individuals are involved and regardless of injuries"[88].

The International Law Commission has long been engaged in a process of codification and progressive development of the basic rules of international law concerning the responsibility of States for their internationally wrongful acts. In this respect, the ILC has crystallized the most fundamental principles of customary international law by drawing upon the holdings of the international courts' and tribunals' most significant decisions. In this respect, according to the ILC draft articles, the concept of international

---

[84] Hackworth, Digest of International Law (1943) quoted by H. M. Kindred, *supra note* 110 at 636.

[85] T. H. Youmans Claim, United States v. Mexico, General Claims Commission (1926), 4 R.I.A.A. 110.

[86] This case involved the laying of mines in Albanian territorial waters. The Court held that every state "has the obligation not to allow knowingly its territory to be used for acts contrary to the rights of other States". Corfu Channel Case (Merits) United Kingdom v. Albania [1949] I.C.J. Rep. 4.

[87] S. Sucharitkul, "State Responsibility and International Liability under International Law" (1996) 18 *Loy. L.A. Int'l & Comp. L.J.* at 826.

[88] *Ibid.* at 827.

responsibility revolves around the idea of Mann's international wrong[89]. In this regard, the ILC understands that "every internationally wrongful act of a State entails the international responsibility of that State"[90]. For the ILC there is an internationally wrongful act of a State[91] when conduct consisting of an action or omission: (a) is attributable to the State under international law; and (b) constitutes a breach of an international obligation of the State[92].

Therefore, the two crucial aspects of international responsibility are the attribution of acts to the State and the breach of an international obligation. With respect to the former, in general a state is responsible for any violation of its obligations under international law when it results from an action or inaction by the government of the state, its political subdivisions or any organ, agency, official or employee acting within the scope of authority[93]. The draft codification of the law of state responsibility of the International Law Commission contains specific norms dealing with the imputability of acts to the State. The primary principle is that the conduct of any State organ[94] is considered an act of that State under international law, whether the organ exercises legislative, executive, judicial or any other functions, whatever position it holds in the organization of the State, and whatever its character as an organ of the central government or of a territorial unit of the State is[95]. Apart from organs of the state, the conduct of persons or entities in the following circumstances are also considered acts of States for the purpose of

---

[89] F. A. Mann coined the term international wrong, giving rise to a unified system of obligations on the international plane. Its theory has rapidly received currency and acceptance by representatives of a wide range of interests in the international community. Under Mann's theory of international wrong when a State denies a remedy it incurs international responsibility. For Mann, the main consequences of an international wrong are the reparation and the nullity of the act. F. A. Mann, Further Studies in International Law (Oxford: Clarendon Press, 1990) at 125.

[90] International Law Commission, Draft articles on Responsibility of States for Internationally Wrongful Acts, 53rd session, 2001, article 1 [hereinafter "ILC International Responsibility"].

[91] The characterization of an act of a State as internationally wrongful is governed by international law. Such characterization is not affected by the characterization of the same act as lawful by internal law. "ILC International Responsibility", *supra note* 90 article 3.

[92] *Ibid.* article 2.

[93] In general, a state is responsible under international law only for official acts, or for official inaction where there was a duty to act. Restatement of the Law, Third, Foreign Relations Law of the United States, § 313.

[94] An organ includes any person or entity which has that status in accordance with the internal law of the State. "ILC International Responsibility", *supra note* 90 article 4.2.

[95] *Ibid.* article 4.

the attribution of international responsibility: (i) the conduct of persons or entities exercising elements of governmental authority[96], (ii) the conduct of organs placed at the disposal of a State, if acting in the exercise of elements of the governmental authority[97], (iii) the conduct of an organ of a State or of a person or entity empowered to exercise elements of the governmental authority if acting in that capacity, even if it exceeds its authority or contravenes instructions[98], (iv) the conduct of a person or group of persons controlled by a State[99], (v) the conduct of a person or group carried out in the absence or default of the official authorities[100], (vi) the conduct of an insurrectional movement which becomes the new government or which succeeds in establishing a new State[101], and (vii) conduct which is acknowledged and adopted by a State as its own[102].

Thus, as arises from the above discussion, ordinarily a State is not responsible for the acts of individuals or other private entities. So, for example, except for the cases dealing with space activities as prescribed by the specific norms dealing with international space responsibility which will be examined below, a state would not be held responsible for damage caused by private persons other than the limited cases described in the ILC draft articles[103].

The other crucial aspect of international responsibility is the question of a breach of an international obligation of the State, whether it arises from custom, treaty or general principle[104]. According to the draft codification of the ILC, there is a breach of an international obligation when an act of a State is not in conformity with what is required by that obligation, regardless of its origin or character[105]. In the Chorzow Factory the Permanent Court of Justice articulated this principle by holding that "it is a principle of international law that the breach of an engagement involves an obligation to make reparation in an adequate form. Reparation is therefore the indispensable complement of a failure to apply a convention and there is no necessity for this to be stated in

---

[96] *Ibid.* article 5.
[97] *Ibid.* article 6.
[98] *Ibid.* article 7.
[99] *Ibid.* article 8.
[100] *Ibid.* article 9.
[101] *Ibid.* article 10.
[102] *Ibid.* article 11.
[103] However, a State may incur responsibility for its failure to provide reasonable protection to foreign individuals or entities.
[104] I. Brownlie, *Principles of Public International Law,* 5th ed. (Oxford: Oxford University Press, 1999) at 439.
[105] "ILC International Responsibility", *supra note* 90 article 12.

the convention itself. Differences relating to reparations, which may be due by reason of failure to apply a convention, are consequently differences relating to its application.[106]"

As arises from the above discussion, objective responsibility constitutes the basis of international responsibility and it has been considered the most effective grounds for upholding the principle of reparation[107]. The majority of the decisions of international courts and tribunals have upheld the objective standard. Thus, for example, the Franco Claims Commission decided that: "the doctrine of the objective responsibility of the State ... [implies] a responsibility for those acts committed by its officials or its organs [....] despite the absence of *faute* on their part."[108]

Unlike most domestic systems of responsibility, in International Law the existence of damage is not a condition for the existence of international responsibility. As put forward by the Permanent Court of Justice in the Chorzow Factory case, it is a principle of international law that the breach of an engagement entails responsibility and involves an obligation to make reparation[109]. Therefore, international responsibility is not "merely a means to allocate risks but, more generally, a tool to enforce standards of conduct imposed on states[110]." Thus, as argued by Judge Krylov in his dissent in the Corfu Channel case "one cannot transfer the theory of risk, which is developed in the domestic law of some States, into the domain of international law."[111]

2. INTERNATIONAL RESPONSIBILITY AND LIABILITY FOR SPACE ACTIVITIES

2. a. International state responsibility

In general, international responsibility for national activities in outer space is inserted within the principles of international state responsibility discussed above. However, with respect to the attribution rules it deviates drastically

---

[106] Chorzow Factory (Indemnity) Case (1928), P.C.I.J., Ser. A., No. 17.
[107] Brownlie, *supra note* 104 at 442.
[108] Caire Claim (1929), RIAAv. 516 at 60, quoted by Brownlie, *supra note* 104 at 440.
[109] Chorzow Factory (Indemnity) Case (1928), P.C.I.J., Ser. A., No. 17.
[110] H. M. Kindred, *International Law Chiefly as Interpreted and Applied in Canada*, 6th edition (Toronto: Edmond Montgomery, 2000) at 636.
[111] Corfu Channel Case (Merits) United Kingdom v. Albania [1949] I.C.J. Rep. 4. Dissent.

from the general norms of international state responsibility. In this regard, article VI of the Outer Space Treaty prescribes that "States Parties to the Treaty shall bear international responsibility for national activities in outer space, including the moon and other celestial bodies, whether such activities are carried on by governmental agencies or by non-governmental entities, and for assuring that national activities are carried out in conformity with the provisions set forth in the present Treaty."[112]. Thus, unlike general International Law the Outer Space Treaty attributes international responsibility to States for national activities in outer space carried on not only by governmental agencies but also by non-governmental entities, i.e., private firms and individuals. This has been considered to be a revolutionary advancement of the doctrine of international state responsibility, for under the attribution rules contained in the Outer Space Treaty the acts and omissions of non governmental entities are considered to be acts imputable to the State as if they were their own acts[113]. As has been put forward by Bin Cheng, international state responsibility in the outer space field arises the moment a breach of an international obligation is produced and not when the State is seen to have failed in its duty to prevent or repress such breach, for the State is immediately accountable for the breach on the international plane as if it itself had breached the international obligation[114].

In light of the above discussion, the question remains as to which state will actually bear international responsibility for their national activities. Thus, it must first be examined what activities are considered national. This issue is directly related to the jurisdiction and the rights of the state to legislate over activities carried out under its colors.

In this context, there is disagreement with respect to the actual notion of national activities among authors and commentators. Two conflicting views have been expressed[115]. One perspective holds that the concept of

---

[112] Outer Space Treaty, article VI.

[113] B. Cheng, Article VI of the 1967 Space Treaty Revisited: "International Responsibility", "National Activities", and "The Appropriate State", (1998) 26 *J.Sp.L.* at 15 [hereinafter "Space Treaty Revisited"].

[114] *Ibid.* at 15.

[115] H. L. van Traa-Engelman, "Problems of State Responsibility in International Space Law", (1983) 26 *IISL* at 140; [hereinafter "Problems of State Responsibility"]; F. v. d. Dunk, "Public Space Law and Private Enterprise. The Fitness of International Space Law Instruments for Private Space Activities", (1998) *IISL* at 4 [hereinafter "Fitness of International Space Law Instruments"]; H. Quizi, "Certain Legal Aspects of Commercialization of Space Activities", (1990) XXII *Ann. Air & Sp. L.* at 337; H. Wassenbergh, *Principles of Outer Space Law in Hindsight* (Dordrecht: M. Nijhoff Publishers, 1991) at 23; B. Cheng "The Commercial Development of Space: the Need

national activities is remitted to domestic law. The other one understands that the Outer Space Treaty provides the content of this notion.

The first perspective may be examined through the vision of one of its main proponents: Henri Wassenbergh. He postulates that the concept of national activities depends on the scope of national legislation. In this respect, Wassenbergh claims that it is each State that should define who their "nationals' are[116]. In this vein, an analogy is drawn with other aspects of international law dealing with nationality. Thus, under this perspective, for example, a state could not consider as its nationals those individual foreign nationals that reside in its territory or those legal entities which are wholly owned by foreign shareholders but that conduct business in its territory on a permanent basis. Thus, for example, that state would not have to authorize and supervise the activities which these individuals and entities carry out in Outer Space. While from a theoretical legal perspective this could be possible in light of the sovereignty doctrine, that exclusion could not have any effect on the international plane, for otherwise, national legislation could be used to circumvent the provisions on international responsibility of states consecrated in the Outer Space Treaty and there may be cases where no state could -deliberately- be considered internationally responsible[117]. Therefore, it is submitted that this cannot be the intention of the drafters of the Outer Space Treaty, for they intended that states bear international responsibility for the activities of their nationals in a very broad sense[118]. Thus, the concept of national activities may not be construed to be at the entire discretion of each state[119].

---

for New Treaties (1991) 19 *J. Sp. L* at 36-40; "Commercial Utilization of Outer Space", *supra* note 2 at 281.

[116] For Wassenbergh, there may be certain cases of conflicts of jurisdiction, which will occur "when the territorial jurisdiction is in the country of residence of such national, i.e., in the foreign country where the space activity takes place, making the activity also a national activity of that country), while the national jurisdiction is in the country of nationality". H. Wassenbergh, "International Space Law: A Turn of the Tide", (1997) XXII *Ann. Air & Sp. L.* at 335 [hereinafter "Turn of the Tide"].

[117] It may be argued that in those cases international responsibility could be attributed to those states that through national legislation refrained from the possibility of authorizing those activities. This argument exceeds the scope of the present discussion, whose purpose is only to interpret the purpose of the treaty clause in question.

[118] A/AC.105/C.2/SR.94 at 51.

[119] "As the considerations to employ a certain criterion and to apply it to a specific case often involve rules of international conflicts law, it needs little imagination to realize the complexity and variety of cases, as well as their outcome according to the different law systems and regulations. If any expectations might be forthcoming from

The second position expressed in the literature regarding the conceptualization of the notion of 'national activities' revolves around the doctrine of jurisdiction and the interpretation of the term "national activities" as contained in article VI of the Outer Space Treaty. In this respect, von der Dunk affirms that the most effective interpretation of [...] 'national activities' may be made in light of the interrelation of the doctrines of jurisdiction and international responsibility[120]. In this sense, it has been put forward that a state may be internationally responsible only for those activities over which it has the possibility to exercise legal control. Thus, national activities within the meaning of article VI of the Outer Space Treaty are those activities over which a state has jurisdiction, or more specifically those activities which it has the possibility to exercise legal control[121]. The opportunity to exercise legal control is encompassed in the concept of jurisdiction[122] and more precisely in the jurisaction of states[123]. Therefore, it is submitted that a State is responsible for the activities over which it has the opportunity to exercise legal control, i.e., activities which are within the state's jurisaction, whether territorial, quasi territorial or personal.

---

such a national approach to serve the aims of international law, and those of space law in particular only an intensified research in the field of comparative law, and a progressive development in the unification of international conflicts law, on a worldwide basis, or at least among space faring nations would have a fair chance of success in establishing international legal security in this respect." "Problems of State Responsibility", *supra note* 115 at 140.

[120] "Fitness of International Space Law Instruments", *supra note* 115 at 4.

[121] *Ibid.* at 4.

[122] "Jurisdiction is a legal concept used to describe a state's right to take action; e.g. to prescribe and enforce rules of law with respect to a particular person, thing, or event. In its inception, the principle of jurisdiction was primarily territorial, deriving from the belief that the power of a nation to act within its own borders was necessarily exclusive and absolute [...] susceptible of no limitation not imposed by itself. Although there are many jurisdictional rationales, all require that there be some genuine link between the state and the person, property, or event over which jurisdiction is claimed". Office of Technology Assessment, "Space Stations and the Law: Selected Legal Issues" in G.H. Reynolds and R.P. Merges, *Outer Space: Problems of Law and Policy*, (Westview Press, 1989) at 248.

[123] Jurisaction is the internationally recognized competence of a state concretely to set up machinery to make, implement and enforce laws, judicial pronouncements and other legally binding decisions. Its counterpart, jurisfaction is the internationally recognized competence of a state to enact laws, make judicial pronouncements and adopt other decisions with legally binding force. "Space Treaty Revisited", *supra note* 113 at 16.

2. b. International State liability

2.b. 1. Genesis of international state liability

International state liability has developed rather autonomously from the doctrine of international responsibility[124]. International liability is based on the proposition that absence of wrongfulness does not preclude the compensation for damage caused by an act of a State[125]. In this respect, it has been argued that:

> "the liability of a State does not stem from its fault or the wrongfulness of its act, but from the injurious consequences suffered by persons beyond its boundaries. [...] International liability prescribes a set of primary rules[126]. In particular, these rules cover situations where the State fails to prevent harmful effects or fails to give necessary warning to avoid and abate such effects. Unlike State responsibility, however, international liability continues to be concerned only with primary rules, while the expanded scope of State responsibility has evolved to deal with breaches of such primary rules or obligations already subsisting in international law, regardless of origin[127]".

The concept of international state liability as distinct from international responsibility has met strong opposition from highly distinguished scholars[128]. For example, Brownlie refuses to consider

---

[124] R. Rosenstock, The Forty-Eighth Session of the International Law Commission" 91 *A.J.I.L.* at 365.

[125] Sucharitkul, *supra note* 87 at 826.

[126] While primary rules refer to the substantive obligations created by international law, e.g., standards for sustainable logging, secondary rules denote the liability incurred as a consequence of breaching those rules. T. A. Berwick, ": Responsibility and Liability for Environmental Damage: A Roadmap for International Environmental Regimes" (1998) 10 *Geo. Int'l Envtl. L. Rev.* at 259.

[127] Sucharitkul, *supra note* 87 at 826.

[128] D. Barstow Magraw, "Transboundary Harm: The International Law Commission's Study of "International Liability" (1986) 80 *A.J.I.L.* at 316.

international liability as an autonomous notion since he understands that much of state responsibility involves lawful activities that have caused harm. Thus, for Brownlie the distinction between lawful and unlawful activities is useless as a relevant factor to resolve the international responsibility and liability debate[129]. Brownlie's position does not find support in international conventional law, for the Outer Space Treaty has explicitly consecrated an international state liability regime, which imposes international liability to the so called launching state regardless of the existence of an internationally wrongful act. Thus, article VII of the Outer Space Treaty prescribes that each State that launches or procures the launching of an object into outer space and each State from whose territory or facility an object is launched, is internationally liable for damage caused by that object to another State or to its natural or juridical persons[130]. However, it must be recognized that the relevance of the notion of wrongfulness is relatively insignificant in the context of absolute and strict liability, such as in the regime created by the Outer Space Treaty[131], since the basis for international liability is the existence of injurious consequences or harm[132].

Some commentators have held that since the Spanish and French language versions of the Outer Space Treaty and the Liability Convention refer only to responsibility, then liability and responsibility are regarded as being one and the same in these languages and in the civil law system[133]. This

---

[129] Brownlie, *supra note* 104 at 461.

[130] Outer Space Treaty, article VII.

[131] This has led several international scholars to center their analysis regarding the difference between international responsibility and liability on the general versus the specific dichotomy of the notion of responsibility and liability respectively. In this respect, Gorove has held that none of these terms have been defined in Space law, but the term "liability" has been used to set the launching state's liability for damage caused by space objects, while the term "responsibility" has been used to mandate international responsibility by the appropriate state for national activities in outer space. Gorove concludes that the term "liability" is linked to the idea of the legal consequences (generally in terms of damages) resulting from a particular behavior. The term "responsibility" refers to obligations imposed on people and institutions who are supposed to carry out or are accountable in certain situations but not necessarily in the form of compensation for damages. S. Gorove, "Liability in Space Law: An Overview" (1983) VIII *Ann. Air & Sp. L.* at 373 [hereinafter "Liability Overview"].

[132] Sucharitkul, *supra note* 87 at 826.

[133] S. Ospina, "International Responsibility and State Liability in an Age of Globalization and Privatization. A Personal View of (established) Texts and (evolving) Contexts" *Fiftieth Anniversary Celebration Institute of Air and Space Law, McGill University*, April 19-21, 2002 [unpublished] at 7.

view neglects to analyze this phenomenon from an international perspective as discussed above. In effect, the system of international state responsibility and liability on the international plane has developed rather autonomously from the domestic systems and it is independent of the constraints and features of its national counterparts. This view, voiced in the legal literature by Syliva Ospina, among others, also fails to take into consideration the fact that both the Spanish and French terms for responsibility, i.e., *responsabilidad* and *responsabilité* respectively, are bi-semic, which encompass both concepts, i.e., the obligation aspect –liability- and the accountability aspect –responsibility-[134] which invalidates this simplified view that assigns "significantly different legal consequences" according to the language of the text in question[135].

Like liability for damage caused by space objects, the international conventions on liability for nuclear incidents also impose international liability on states. However, the state has only a subsidiary role as liability for nuclear damage falls primarily on the operator of the nuclear installation and the installation State, i.e., the state that licenses the nuclear operator, is only liable for damages which exceed the amount of insurance up to the limit of the operator's liability[136].

In the space arena the principles of international state liability have been further developed in the Liability Convention. Created at the peek of the cold war, the main objective of the Liability Convention was to provide a legal framework for the full compensation of damage caused on Earth by the spacefarers as a result of their activities in outer space[137]. According to Maniatis, the Convention is based on two factual premises: (i) the greatest risks posed by space activities are to non-participants, and (ii) States are the principal space participants. Therefore, the system deliberately ignores the doctrine of private liability and allocates responsibility for damage caused by space objects by resorting to state imputability rules.[138]

---

[134] P. Hannapel, (chair) "Privatization of Space Activities" *Fiftieth Anniversary Celebration Institute of Air and Space Law, McGill University*, April 19-21, 2002 [unpublished].

[135]Ospina, *supra note* 133 at 7.

[136] S. D. Murphy, "*Prospective Liability Regimes for the Transboundary Movement of Hazardous Wastes"* (1994) 88 *A.J.I.L.* at 24.

[137] "Commercial Space" *supra* note 7 at 62.

[138]D. Maniatis, "The Law Governing Liability for Damage Caused by Space Objects: From State Responsibility to Private Liability" (1997) XXII *Ann. Air & Sp. L.* at 373. With the exception of articles dealing with ratification, amendments, withdrawal and review of the Convention, references in the Convention to states also

2) b. 2. Relationship between private entities and international state liability

The Liability Convention does not in itself, at least explicitly, attribute international liability to the launching state for damage caused by non governmental entities. However, the interplay between the responsibility and liability provisions of the Outer Space Treaty and the Liability Convention leads to the unquestionable conclusion that under the Liability Convention, States are liable for damage caused by the space objects of their national private entities, since, as arises from the above discussion, the Outer Space Treaty imposes international responsibility to States for the national activities in outer space carried out by non-governmental entities[139] and relates the article VII international state liability principle to the principle of international state responsibility[140].

However, it has been suggested that this conclusion may not be valid in the cases of States that are party to the Liability Convention but are not parties to the Outer Space Treaty[141]. This proposition neglects to consider the validity of customary international rules in the governance of outer space activities[142]. In effect, ever since the beginning of the Space Age, space

---

applies to any international intergovernmental organization which conducts space activities if the organization declares its acceptance of the rights and obligations in the Convention and if a majority of its members are parties to the Convention as well as the Outer Space Treaty. All such state members are to take all appropriate steps to ensure that the organization makes such a declaration. If an organization is liable under the Convention, it and its members who are parties are jointly and severally liable provided that any claim is first presented to the organization and it has not paid within six months the amount agreed or determined to be due. Any claim on behalf of an organization which has made such a declaration is to be presented by a member which is also a party to the Convention. There is no principle or rule of international law whereby a group of states can, by international agreement, impose legal obligations on an international organization without the acquiescence of that organization. Although the Outer Space Treaty has provisions to make it applicable to international organizations, such entities would not be legally bound thereby without their acceptance thereof. Committee on Aeronautical and Space Sciences United States Senate, Convention on International Liability for Damage Caused by Space Objects. Analysis and Background Data, 92d Congress 2d. Session, US. Government Printing Office, Washington, 1972 at 39 [hereinafter "Committee Report"].

139 Outer Space Treaty, article VI.

140 Outer Space Treaty, article VII.

141 W. F. Foster, "The Convention on International Liability for Damage Caused by Space Objects" (1972) *Canadian Yearbook of International Law* at 137.

142 Restatement of the Law, Third, Foreign Relations Law of the United States, § 102.

activities were conducted in accordance with international law, which includes customary international rules, as well as general principles[143]. Furthermore, a series of General Assembly resolutions reiterated the applicability of customary international law to the outer space arena and the Outer Space Treaty expressly refers to the applicability of international law, which again embraces customary international law as well as general principles[144]. Therefore, since the attribution of international responsibility to States for the national activities of their non governmental entities is a rule of customary international law the concern raised with respect to the possibility of the existence of States that are parties to the Liability Convention but not to the Outer Space Treaty is fallacious and thus damage caused by private or other non governmental firms would trigger off international liability of the respective launching states even for those States which have not ratified or acceded to the Outer Space Treaty[145].

Some very respected publicists have also arrived at the conclusion that damage caused by private entities generates international state liability on different legal grounds. For example, Böckstiegel[146] and Cheng[147] hold that the liability provisions are applicable both for launchings by states and by non governmental institutions because invariably at least one of the four criteria for the launching state will be fulfilled in the case of launchings by private entities. This reasoning is erroneous first because it ignores the clear provisions of the Outer Space Treaty[148] and the norms of customary international law and second because there may be cases where a private company may launch a payload without the active involvement of any state,

---

[143] B. Cheng, "The Commercial Development of Space: The Need for New Treaties" (1991) *J. Sp. L.* at 19.
[144] Maniatis, *supra note* 138 at 373.
[145] The substantive provisions of the Convention have been formulated in general terms and not expressly stated to be applicable only as among contracting parties. In effect, the Convention reflects statements of general international law of universal application, independently of the treaty, which has led Bin Cheng to characterize its application as ecumenical. Furthermore, Cheng understands that there is support in the Convention for procedural provisions such as the presentation of claims and the Claims Commission to be binding upon non-contracting parties. B. Cheng, "Convention on International Liability for Damage Caused by Space Objects" in N. Jasentuliyana & R.S.K. Lee eds., *Manual on Space Law* (New York: Oceana, 1979) Vol. 1 at 98 [hereinafter "Manual on Space Law"].
[146] K. H. Böckstiegel, "The Term Launching State in International Law" (1994) 37 *IISL* at 81.
[147] B. Cheng, "Convention on International Liability for Damage Caused by Space Objects", "Manual on Space Law", *supra* note 145 at 103.
[148] Outer Space Treaty, article VI and VII.

such as the case of a launch from a private launch facility located outside the territory of a State where a State neither procures nor launches a space object[149].

2) b. 3. The launching state

The Liability Convention attributes international liability to the launching state, which is defined as "a state which launches or procures the launching of a space object, or a State from whose territory or facility a space object is launched[150]." This article shows the complexity of the launch of space objects and the myriad of States which may be involved in a launching operation[151]. The definition of the concept of launching state, which mirrors the standards of article VII of the Outer Space Treaty and the definition contained in the Registration Convention, provides sufficient basis for the determination of the State which bears international liability. However, this definition has given rise to some concerns in the legal literature, especially around the concept of procuring state [152]. In this respect, for example, Carl Q. Christol wonders exactly what degree of activity qualifies a procuring state as such[153]. In the same line of reasoning, for the US Senate Committee on Aeronautical and Space Sciences it is not clear in the Convention whether a State would fall within the category of procuring State if its only connection with a space activity is a minor experiment aboard the spacecraft, or if it supplied only a small component in the spacecraft booster or if it just sent a technical observer[154]. Christol concludes that this question has been left open in the Convention[155]. Böckstiegel suggests that there should be a substantial threshold test and therefore the provision of small minor components to the payload or the launching of another State, and even the sale of a satellite to another State would not qualify as procurement. However, as Böckstiegel

---

[149] Von der Dunk's interpretation is analogous to those of Bin Cheng and Böckstiegel but it differs in the fact that he articulates his thesis in a potential mode, thus leaving room for cases where there may be no involvement of states in the launching operations. F. G. von der Dunk, "Launching from Down Under: The New Australian Space Activities Act of 1998" (2000) (43) *IISL* at 134.

[150] The Convention also extends the meaning of the term "launching" to include attempted launchings. Liability Convention, article I.

[151] P. Nesgos, "The Challenges Facing the Private Practitioner: Liability and Insurance Issues in Commercial Space Transportation" (1989) 4 *J.L. & TECH.* at 23.

[152] See "Commercial Space" *supra* note 7 at 70.

[153] C. Q. Christol, *The Modern International Law of Outer Space*, (New York: Pergamon Press, 1982) at 115.

[154] "Committee Report", *supra* note 138 at 29.

[155] Christol, *supra* note 153 at 115.

himself recognizes there are in practice many situations which are not at all clear. Thus, it is submitted that each decision as to whether a State falls within the category of procuring state is a question of fact, which should be made on a case by case basis in light of the parameters contained in the definition of launching state. In this respect, Christol's assertion that the Convention has purposefully been left open supports our conclusion that the qualification of a State as a launching state can only be decided in each specific case of damage arising from a space endeavor.

Another deficiency concerning the delimitation of the concept of launching state arising from the text of the Convention is found in article V, which attributes joint liability for any damage caused by two or more States when they jointly launch a space object[156]. Article V neglected to include the procuring State among those which may be jointly liable. As arises from our above discussion, the definition of article I, which includes the procuring State, together with the general principles of joint liability established in the Convention[157] leads to the unquestionable conclusion that a procuring state is to be regarded as a participant in a joint launching, and thus it is subject to joint liability in terms of article V of the Convention. Furthermore, it has been argued that the purpose of establishing several launching states is to ensure that the victim has ample possibilities to be compensated[158]. Therefore a literal reading of article V as excluding the procuring state within those States that must bear joint liability for damage would run contrary to the purpose of the whole Convention.

As anticipated above, the other categories of launching states, i.e., the State which launches and the State from whose territory or facilities a space object is launched, are more straightforward and present fewer possibilities of ambiguities. Nonetheless, some controversy has arisen with respect to the launches from the sea, such as the case of the Sea Launch company[159]. There are additional difficulties which may arise from situations which have not

---

[156] Outer Space Treaty, article V. It is possible under the Convention for participants in a joint launching to conclude agreements regarding the apportioning among themselves of the financial obligation in respect of which they are jointly liable. These agreements however may not impair the right of a state sustaining damage to seek the entire compensation due from any or all of the launching States. Also, in the absence of said agreement, the State which has paid compensation for damage is entitled to present a claim for indemnification to other participants in the joint launching.

[157] See "Commercial Space" *supra* note 7 at 68.

[158] "Space Treaty Revisited", *supra note* 113 at 15.

[159] M. Longo, "Legal Aspects of Launching Space Objects from Non-Terrestrial Sites" (1999) 42 *IISL* at 323.

been expressly contemplated in the definition of launching state. For example, a problem may arise from the sale of a satellite in orbit. For Kerrest, in case of the sale of a satellite to a national of a State which was not an original launching state, this new State may not be held liable under the Outer Space Treaty and the Liability Convention, but it could be under general International Law[160]. In light of our above discussion on the notion of international state responsibility and the concept of national activities under the Outer Space Treaty it may not be concluded but to affirm that the new State will be internationally responsible, since the use of a satellite in orbit will definitely qualify as a national activity of the new State, as it will have the opportunity to exercise legal control, i.e., the use of a satellite in orbit will be within the new state's personal jurisaction, which undoubtedly entails its responsibility on the international plane under the Outer Space Treaty[161]. The State whose national sells the satellite would continue to be a launching state under the Liability Convention, for the Convention does not foresee the possibility of extinguishing liability in any circumstance[162]. However, the new State can execute an agreement with the state whose national sold the satellite whereby the former assumes all liability which may arise from damage caused by the satellite after its sale and consequent transfer of title and whereby it holds the latter harmless and agrees to indemnify it for any loss which it may incur. The Liability Convention itself would allow this possibility, since article V authorizes the possibility of agreements to allocate the financial obligation among States[163]. This, however, will not have effect vis-à-vis the victim whose national State could always seek the entire compensation from any launching States, including the State of the seller of the satellite[164].

2) b. 4. Liability standard and damages

The Liability Convention adopted an absolute liability standard, i.e., objective liability, where the victim does not have to prove the defendant's fault, without any monetary limits, for damages caused by a space object on the surface of the earth or to an aircraft in flight[165]. This parallels the absolute

---

[160] Kerrest, "Remarks on the Notion of Launching State" (1999) 41 *IISL* at 309 [hereinafter "1999 Remarks"]

[161] See *supra* note 123.

[162] R. J. Lee, "Effects of Satellite Ownership Transfers on the Liability of the Launching States" (2000) 43 *IISL* at 148.

[163] Liability Convention article V.2.

[164] *Ibid.* article V.2.

[165] However, the Convention foresees the possibility of the exoneration from absolute liability. In effect, article VI .1 "establishes that exoneration from absolute liability shall be granted to the extent that a launching State establishes that the damage has

liability standard contained in the four international conventions on liability for nuclear incidents[166]. These conventions consecrated the principle of absolute liability for nuclear hazards and its exclusive imposition on the operators of nuclear installations under four separate international legal frameworks for a civil action for indemnity on the basis of the domestic law of individual contracting parties. However, unlike the Liability Convention they also adopt international minimum and maximum levels for compensation[167].

Additionally, for damages which take place elsewhere than on the surface of the earth by (i) a space object of a launching State, and (ii) persons or property on board such a space object, the Liability Convention adopted a subjective standard, where evidence of negligence is required (article III). As in the case of objective liability, article III claims are not subject to any monetary limitations[168].

---

resulted either wholly or partially from gross negligence or from an act or omission done with intent to cause damage on the part of a claimant State or of natural or juridical persons it represents". Paragraph 2 of the quoted article sets forth that "no exoneration whatever shall be granted in cases where the damage has resulted from activities conducted by a launching State which are not in conformity with international law including, in particular, the Charter of the United Nations and the Treaty on Principles Governing the Activities of States in the Exploration and Use of Outer Space, including the Moon and Other Celestial Bodies". Exoneration under the Convention may be total or partial depending on the participation of the claiming State. Forkosch questions the fact that there is no standard to determine when negligence should be considered gross and wonders how an omission could be considered to be "with intent" to cause damage. In our opinion, these issues are not exclusive of Space Law and have satisfactorily been answered in other fields of law. M. D. Forkosch, *Outer Space and Legal Liability* (Dordrecht: Martinus Nijhoff, 1982) at 48.

[166] These are the 1960 Paris (OECD-NEA) Convention, the 1963 Vienna Convention, together with their respective supplementary protocols, the 1962 Brussels Convention concerning (merchant) ships with nuclear propulsion; and the 1971 Brussels Convention on liability for ocean transportation of nuclear substances. P. Lang, "La responsabilité internationale encourue en raison des activités liées a l'utilisation de l'energie nucleaire by Angelo Miatello. BOOK REVIEW", (1989) 83 *A.J.I.L.* at 965.

[167] P. Lang, "La responsabilité internationale encourue en raison des activités liées a l'utilisation de l'energie nucleaire by Angelo Miatello. BOOK REVIEW", (1989) 83 *A.J.I.L.* at 965.

[168] It is under discussion whether the attribution of liability based on fault is applied only in outer space and celestial bodies, or also in cases of damage caused by a space object to persons or goods other than aircrafts in airspace. We understand that liability based on fault is an exception to the absolute liability standard, and as such its

The core of the Liability Convention is the full compensation standard imposed on the launching state, which has to restore the victim to the condition which would have existed if the damage had not occurred.[169] This principle, known as full compensation or *restitutio in integrum*, has been borrowed from the Permanent Court of Justice's decision in the Chorzow Factory where the Court held: "that reparation must, as far as possible, wipe out all the consequences of the illegal act and reestablish the situation which would, in all probability, have existed if that act had not been committed.[170]"

The Convention has not adopted any domestic law to govern the recovery of damage[171]. Rather, it has opted for International Law and the principles of justice and equity[172], which solves the problem that may arise from the fact that damage is not equally compensated in every domestic law system.[173]

---

application is restricted only to the cases specifically stipulated in article III of the Convention.

[169]Proposals have been made to advance from the system of absolute liability towards total responsibility. While the former leads to the mere compensation of damages, the latter implies a double penalty, both economic and juridical, because of the deep ethical contents it entails. A. A. Cocca, "From Full Compensation to Total Responsibility", (1983) 26 *IISL* at 157.

[170] Chorzow Factory (Indemnity) Case (Germany v. Poland), P.C.I.J., ser. A, No. 17, at 47 (1928). The Court added that this result can be accomplished in several ways: through restitution in kind; or, if this is not possible, through "payment of a sum corresponding to the value which a restitution in kind would bear"; and "the award, if need be, of damages for loss sustained which would not be covered by restitution in kind or payment in place of it."

[171] It has been defined as loss of life, personal injury or other impairment of health, or loss of or damage to property of states or of persons, natural or juridical, or property of international intergovernmental organizations.

[172]This led Carl Q. Christol to affirm that "the virtue of Article 12 was that it avoided a multiplicity of inconsistent and conflicting domestic outlooks in favor of a coherent and cohesive international standard." Christol, *supra* note 153 at 116.

[173] It is worth noting that article X prescribes that a claim for compensation for damage may be presented to a launching State no later than one year following the date of the occurrence of the damage or the identification of the launching State which is liable. The second paragraph of said article introduces the possibility of presenting a claim beyond the one-year period if a state does not know of the occurrence of the damage or has not been able to identify the launching state. In said circumstances the claim may be presented within one year following the date on which the state learned of the aforementioned facts; however, this period may in no event exceed one year following the date on which the state could reasonably be expected to have learned of the facts through the exercise of due diligence. Finally, the last paragraph allows the presentation of the claim even beyond such periods

2) b. 5. Specific liability arrangements

The Intergovernmental Agreement on the International Space Station contemplates a special regime for the allocation of liability which includes liability arising under the 1972 Convention. Its objective is to establish a cross-waiver of liability by the Partner States and related entities with the purpose of encouraging participation in the exploration, exploitation, and use of outer space through the Space Station[174]. Cross waivers of liability originated in the first launch services agreements executed by NASA[175], which were later adopted by all major launch carriers around the world.[176] They constitute the milestone of any space risk management system[177] and are generally complemented by other space risk management tools[178], which makes the risk allocation and assignment of liability in the space field a complex system with well-defined characteristics[179]. By means of these waivers of liability, each party agrees to be responsible for any damage which it sustains as a result of damage to its own property and employees, whether

---

when the full extent of the damage is not known. In such event the claimant state is entitled to revise the claim and submit additional documentation after the expiration of such time-limits until one year after the full extent of the damage is known. In other words, partial claim of damages is permitted, provided the full extent of damages is unknown. Thus a claim may be filed each time new damages arise from the same event. Liability Convention, article X.

[174] International Space Station Intergovernmental Agreement, article 16.

[175] 14 Code of Federal Regulations Chapter V § 1214.113. This reads as follows "The U.S. Government will assume no risk for damages to the customer resulting from certain activities conducted under the launch agreement or to third parties resulting from launch related or on- orbit operations. The customer will be required to agree to be bound by a cross-waiver of liability among the customer, other customers, related entities and NASA for all activities under the launch agreement. The customer will also be required to purchase third-party liability insurance covering launch and on-orbit operations in an amount deemed appropriate by NASA." Notice that the regulations do not determine the scope, the extent, conditions or duration of the reciprocal waivers of liability. It only mentions the general purposes sought by the use of this legal instrument. Therefore, the waivers have their actual regulation in the launch services agreements. In effect, as recognized by NASA in its Launch Service Agreement these liability risks are distributed between the customer and the carrier in accordance with the contract specifications.

[176] "Martin Marietta Case", *supra* note 66 at 239.

[177] "Space Risk Management" *supra* note 60 at 9.

[178] These include indemnification granted by the states, commitments to obtain insurance, limitations of liability, sole contractual remedies in the event of default, *obligations de moyens*, and exclusion of liability clauses, among other legal instruments. "Space Risk Management" *supra* note 60 at 10.

[179] "Space Risk Management" *supra* note 60 at 9.

the damage is caused by the carrier, the customer or other customers involved in the space transport operations and waives all claims against the other parties. As is the case in the International Space Station Agreement, usually, this is complemented by the obligation imposed on all parties to the agreement to include similar waivers of liability in their agreements with other related entities, so that each will assume its risks and will not sue the other participants[180].

As arises from the above discussion, these waivers of liability consist of (i) a general assumption of risks by each party, (ii) the assumption of the consequences of those risks, (iii) a consequent waiver of rights to make a claim for liability, and (iv) an indemnification or hold harmless provision in case of actions filed despite the waiver.[181] The purpose of the reciprocal waivers of liability is twofold: first to limit the claims that might arise from a launch, and second to minimize the need to obtain insurance to protect against claims which may otherwise derive from the launch. In effect, under a reciprocal waiver of liability a party is precluded from making a claim, whether judicial, administrative or otherwise, to the other party or parties to the reciprocal waiver of liability agreement[182].

It has been suggested that the risk allocation regime established under the International Space Station Agreement constitutes an exception to the liability regime consecrated by the Liability Convention[183]. As arises from the above discussion, the Liability Convention allows the possibility of arrangements between launching states to distribute the risks arising from a joint launch[184]. These agreements, however, may not impair the right of a non participant state sustaining damage to seek the entire compensation due from any or all of the launching States. It is thus submitted that the risk distribution regime of the International Space Station agreement qualifies as an agreement among launching states to redistribute their financial obligations in terms of

---

[180] In the International Space Station Agreement, the term related entity includes: (1) a contractor or subcontractor of a Partner State at any tier; (2) a user or customer of a Partner State at any tier; or (3) a contractor or subcontractor of a user or customer of a Partner State at any tier.

[181] The cross-waiver of liability does not apply to: (1) claims between a Partner State and its related entity or between its own related entities; (2) claims made by a natural person, his/her estate, survivors or subrogees (except when a subrogee is a Partner State) for bodily injury to, or other impairment of health of, or death of such natural person; (3) claims for damage caused by willful misconduct; and (4) intellectual property claims. IGA, article 16.

[182] "Space Risk Management" *supra* note 60 at 109.

[183] See *supra* note 174.

[184] Liability Convention, article V.2.

article V of the Liability Convention. These agreements are valid only among these States and are not opposable to non participating states. Furthermore, article XXIII of the Convention supports this conclusion, as it further prescribes that the Convention has no effect on other treaties so far as relations between parties are concerned and that states can enter into treaties reaffirming, supplementing or extending its provisions, provided, however, that these agreements do not affect the rights of the victims.[185]

2) b. 6. Non applicability

The Liability Convention does not apply to: (i) nationals of the launching state, and (ii) foreign nationals who participated in the operation of that space object. According to Bin Cheng, the first exception is an application of a basic principle of International Law which refrains from dealing with relations between a state and its nationals, and the second one is an application of the principle *Volenti non fit jura*[186]. As stated, by Herbert Reis, US representative before COPUOS in the 1967 session, the second exclusion was designed to exempt the launching state from liability for foreign observers who accepted invitations to take part in or observe a launching or recovery since these persons could be considered to have assumed any risk entailed. Nonetheless, according to Reis this exclusion does not imply that the launching state might not pay compensation: it might be paid, for example, under article VII of the Outer Space Treaty.[187]

2) b. 7. Procedural issues

The main obstacle for the private space sector with respect to the dispute resolution regime consecrated in the Liability Convention lies in the fact that it does not allow private companies to resort to the dispute settlement procedure directly but through the States. Additionally, this regime has raised

---

[185]In effect, existing or subsequent treaties, whether bilateral or multilateral, which may cover the same subject, or any Outer Space activity, are not affected by the Convention as to the rights between the parties spelled out therein. For example, according to the Committee on Aeronautical and Space Sciences United States Senate, the Convention would have no effect on the relationship of liability between the state parties to INTELSAT Agreement, which provides that neither INTELSAT nor any of the signatory states, nor any employee or agent acting within their scope of authority will be liable to one another for the unavailability, delay or faultiness of telecommunications services. "Committee Report", *supra note* 138 at 40.

[186]B. Cheng, "Convention on International Liability for Damage Caused by Space Objects", "Manual on Space Law", *supra note* 145 at 103.

[187]A/AC.105/C.2/SR.77 at 5.

unanimous criticism and has even led some States, such as Canada, to abstain from voting for the Convention in the United Nations General Assembly. Most of the criticism centers around the fact that the decisions of the Claims Commission are not binding. These issues demand an examination of the regime.

The first method to resolve a dispute between states is negotiations through diplomatic channels. In the event that a settlement is not reached within one year[188], the parties concerned may establish a Claims Commission. It is to be composed of three members, one appointed by the claimant state, another appointed by the launching state. The third member, who will act as the chairman, is to be chosen by both parties or by the Secretary-General of the United Nations in the event claimant and defendant may not reach an agreement. As anticipated above, one of the weakest aspects of the Convention is the fact that the awards of the Claims Commission are not legally binding. In effect, the decision of the Commission is final and binding only if the parties have so agreed. Otherwise, the Commission only renders a final and recommendatory award, which the parties are to consider in good faith. However, it must be noted that as the result of the obligation to consider the award in good faith, introduced in the text of the Convention by the Argentine representative, the decision of the Claims Commission is not altogether empty of content[189]. Additionally, several authors and commentators have advocated for the amendment of the Liability Convention to make the awards of the Claims Commission binding. This suggestion is politically *naïve*, for an international convention is not changed because of scholarly concerns but for political interests of the main actors of the international community and momentum for such change is not likely to be gained in the near future. Thus, the position adopted by the International Law Commission, led by Maureen Williams -the current chair of its Space Law Committee- seems preferable. In this respect, Williams has resorted to encouraging States to make the option included in paragraph 3 of the General Assembly's Resolution 2777 (XXVI) for the decisions of the Claims Commission to be binding. This same recommendation was made by Austria at the Legal Subcommittee of COPUOS in 1998[190].

Unlike other fields of international law, the Convention does not require the prior exhaustion of any local remedies for the presentation of a claim for compensation to a launching State. Additionally, there is nothing in

---

188 This is counted as from the date on which the claimant state notifies the launching State that it has submitted the claim.

189 "Commercial Space", *supra* note 7 at 77.

190 "ILA Report", *supra* note 82 at 16.

the Convention that prevents a claimant from pursuing a claim in the courts or administrative tribunals or agencies of a launching state. It is thus necessary to inquire which domestic laws would be applicable. While every nation has its own methods for choosing the law applicable to a specific case, the following methods are the most common ones: (i) *lex loci delicti*, i.e., the law of the place where the tort occurred, this could be the law of the State which has jurisdiction on a space object pursuant to article VIII of the Space Treaty, (ii) *lex fori*, i.e., the law of the forum where the case is brought, and (iii) the law of the State having the greatest interest, i.e., the law of the State with the closest connection to the incident.[191]

Apart from the non binding character of the awards of the Claims Commission, the Liability Convention seems inadequate to govern present commercial space activities carried out by private entities, for it does not permit the direct presentation of claims by non governmental entities[192]. In effect, the only entities which may make a claim under the Liability Convention are the States. Furthermore, States are not even obliged to make a claim whenever they or their nationals, whether individuals or private companies, suffer damage. Therefore, article VIII of the Liability Convention permits States other than the State of nationality to make a claim. In this respect, it establishes that if the State of nationality does not present a claim, the State in whose territory the damage was sustained may do so. Otherwise, the State of permanent residency of the victims may act[193].

Therefore, private companies may only make a claim to a foreign State by requesting the State of their nationality or the other States in the cases contemplated in article VIII to present a claim. However, none of these States are obliged to make the claims. With respect to the State of nationality, absent a specific domestic measure for the presentation of claims, whether general or specific for damage caused by space objects, the presentation of a claim will follow the rules for the espousal of claims under general rules of Public International Law and those specific provisions of the Liability Convention[194]. As arises from the Mavrommatis Palestine Concessions case between Greece and the United Kingdom "it is an elementary principle of

---

[191] Reynolds & Merges, *supra* note 122 at 299.

[192] T. Ballarino & S. Busti, *Diritto Aeronautico e Spaziale* (Milano: Giuffrè, Milano, 1988) at 175.

[193] Liability Convention, article VIII.

[194] General rules of International Law must be interpreted in the context of the Liability Convention. This implies, for example, that the prior exhaustion of local remedies mandated under International Law does not apply, for the Liability Convention specifically waives this requirement.

international law that a State is entitled to protect its subjects, when injured by acts contrary to international law committed by another State, from whom they have been unable to obtain satisfaction through the ordinary channels. By taking up the case of one of its subjects and by resorting to diplomatic action or international judicial proceedings on his behalf, a State is in reality asserting its own rights –its right to ensure, in the person of its subjects, respect for the rules of international law.[195]" It is important to notice that under International Law States are not obliged to espouse claims and thus their decision is thoroughly discretionary. In Canada, for example, the Department of External Affairs issued guidelines whereby it will only "espouse claims in respect of loss of human life, property, rights, interests or debts of [...] Canadian citizens. [...] As regards claims by companies, the Government of Canada, pursuant to customary international law as interpreted in the Barcelona Traction case, may espouse claims in respect of property nationalized or otherwise taken abroad where the claims belong to a company incorporated under the laws of Canada or any province of Canada [...] There is a further requirement in Canadian practice that company claims will normally only be espoused by the Government of Canada where there is a substantial Canadian interest in the company. [196]"

The only country which has enacted a specific law to deal with the specificity of the problems arisen from damage caused to individuals and non governmental entities by space objects is Italy, that, on January 25, 1983, enacted a law[197] whereby the Italian State is obliged to compensate the

---

[195] Mavrommatis Palestine Concessions Case, Greece v. United Kingdom (1924), P.C.I.J., Ser. A., No.2. Under customary international law, the espousal of claims is only possible in the case of continuous nationality, i.e., the claimant had to be a national of that state at the time when the claim arose and continuously thereafter until the claim is presented. M. N. Leich, "U.S. Practice" (1982) 76 *A.J.I.L.* at 836. In the words of the Permanent Court of Justice: "this right is necessarily limited to intervention on behalf of its own nationals because, in the absence of a special agreement, it is the bond of nationality between the State and the individual which alone confers upon the State the right of diplomatic protection, and it is as a part of the function of diplomatic protection that the right to take up a claim and to ensure respect for the rules of international law must be envisaged. Where the injury was done to the national of some other State, no claim to which such injury may give rise falls within the scope of the diplomatic protection which a State is entitled to afford nor can it give rise to a claim which that State is entitled to espouse." Panevezys-Saldutiskis Railway, Judgment, 1939, P.C.I.J., Series A/B, No. 76.

[196] J. G. Castel, Legal Services provided by the Department of External Affairs with respect to International Judicial Co-operation and Other Matters, quoted by H. M. Kindred, Kindred, *supra* note 110 at 674.

[197] Italian Law, January 25, 1983.

damage caused by foreign space objects when the Italian State has requested and obtained the compensation of damage from the launching state. However, the law also permits the compensation of damages to Italian citizens even if Italy has not actually obtained the compensation of such damages from the launching state[198]. Additionally, this law contemplates the compensation of damages to foreigners when Italy acts as the territorial state[199] of the State of permanent residents of such foreigners[200]. This law clearly solves most of the problems which private companies may face when they sustain damage caused by a space object. Furthermore, the Italian law relieves private companies from the burdensome task of having to resort to the Public International Law rules regarding espousal of claims.

Thus, absent a law as the one adopted by Italy applicable also to private entities, in practice the fact that private companies may not directly make a claim against a launching state, together with these restrictions and qualifications of the doctrine of the espousal of claims arising from customary international law, will make it burdensome for private companies to obtain remedy in an expeditious manner for loss of property arisen from damage caused by space objects. It is thus recommended that States adopt a specific norm to deal with these difficulties based on the Italian model.

3. CONCLUDING REMARKS ON INTERNATIONAL RESPONSIBILITY AND LIABILITY

As arises from the foregoing discussion, international responsibility has evolved as customary international law, and is centered around the concept of wrongful act. Its two crucial aspects are the attribution of acts to the State and the breach of an international obligation. Generally a state is not responsible for the acts of individuals or other private entities.

Considered the most effective grounds for upholding the principle of reparation, objective responsibility constitutes the basis of international responsibility. In International Law the existence of damage is not a condition for the existence of international responsibility and thus it acts as a tool to enforce standards of conduct imposed on States rather than as a means to allocate risks. The regime of international responsibility for space activities departs radically from the general norms of international state responsibility with respect to attribution rules, for States have been assigned international

---

[198]Ballarino, *supra* note 192 at 175.
[199] Italian Law on Space Responsibility, article VIII.2
[200] *Ibid.* article VIII.3.

responsibility for national activities in outer space carried on not only by governmental agencies but also by non-governmental entities, i.e., private firms and individuals. In this respect, a State is responsible for the activities over which it has the opportunity to exercise legal control, i.e., activities which are within the state's jurisaction, whether territorial, quasi territorial or personal.

The Liability Convention of 1972 has been elaborated upon the principles of international state liability. Although it does not explicitly attribute international liability to the launching state for damage caused by private entities and individuals, States are liable for damage caused by its national private entities on account of the responsibility and liability provisions consecrated in the Outer Space Treaty.

The Liability Convention has structured a dual objective and subjective system of unlimited state liability, which implies an onerous burden for States as it is considered that liability arising from space activities is the most omni-comprehensive liability regime. Therefore, it is concluded that at the national level States must adopt a regime to protect themselves from the consequences of these activities. This protection should be twofold. First, States should structure safety laws or other measures to minimize the risks derived from the space endeavors of their nationals. Second, States should adopt national legislation or other domestic legal measures establishing that States may recover all or part of the compensation paid at the international level from the actual doer of the wrong, for otherwise, this international liability system also implies the assumption of risks and liability of non governmental entities by the States themselves. Therefore, it is submitted in the present study that States need to adopt a risk distribution system to reallocate these risks and liability according to their political objectives in the space arena.

B. AUTHORIZATION AND SUPERVISION

1. INTRODUCTORY ASPECTS

Article VI of the Outer Space Treaty[201] provides that the activities of non-governmental entities in outer space, including the moon and other celestial bodies, will require authorization and continuing supervision by the appropriate state[202]. First, it is necessary to consider whether or not the principle of state authorization and continuing supervision creates an obligation to enact domestic space legislation.

Then we will examine the content and extent of the authorization and supervision principle. This will determine the faculties and bases of States to implement the authorization and supervision obligations in their own jurisdictions whether they do so through the enactment of special domestic legislation or not. The starting point for this analysis is the examination of the language of the principle of authorization and continuous supervision[203] and the ensuing legal lacunae which it creates. Then, consideration will be given to the interpretation of this principle within the framework of the *Corpus Juris Spatialis* in general and the Outer Space Treaty in particular. This will permit us to determine the limits of the states for the domestic implementation of their obligations[204].

2. THE ENACTMENT OF DOMESTIC LAW

It has been argued that the Outer Space Treaty establishes an obligation to pass domestic legislation in every state to enforce the supervision and

---

[201] Outer Space Treaty, article VI. "The activities of non-governmental entities in outer space, including the moon and other celestial bodies, shall require authorization and continuing supervision by the appropriate State Party to the Treaty."

[202] The authorization requirement parallels the license requirement contained in the International Telecommunication Union conventions. The ITU Convention and Constitution recognizes that States are sovereign to regulate their telecommunications. This right has been translated as license requirements to provide telecommunications services. Preambule to the ITU Constitution and Convention of the International Telecommunication Union, 22 December 1992 (Geneva: ITU, 1992) preamble.

[203] Outer Space Treaty, Article VI.

[204] We will try to answer the following questions: what should be authorized and controlled? When? and who should authorize and control?

authorization requirement[205]. In this respect, it has been held that all States party to this Treaty "have the obligation to adopt national legislation whose object shall be the organization of the conditions of the issuance of the authorization to exercise space activities, together with the form of control which they must assure on a continuous basis.[206]" Nonetheless, proponents of this argument understand that States are the only judges of the moment when they will fulfill this obligation[207] and the form and content of that national legislation[208].

However, an analysis of the language of the Treaty[209] reveals that there is no single provision that actually mandates the enactment of national law. As pointed out by Valérie Kayser,[210] the only obligation for States contained in the Outer Space Treaty consists of authorizing and supervising the activities of non governmental entities and to be internationally responsible for the possible damages caused by them. Kayser further holds that the obligation to enact domestic legislation "is not provided for in any provision of the Outer Space Treaty. The Treaty does not say that States 'shall' enact legislation to fulfill their obligations. All its says is that the activities of non governmental entities in outer space [...] shall require authorization and continuing supervision by the appropriate State Party to the Treaty"[211].

A close examination of the genesis of this principle will help to clarify the present discussion. The principle of authorization and continuous supervision was first incorporated into an international instrument, with

---

[205]M. Bourély, "Quelques réflexions au sujet des législations spatiales nationales" (1991) XVI *Ann. Air & Sp. L.* at 247 [hereinafter "Législations spatiales nationales"].
[206] *Ibid.* at 249.
[207] Bourély recognizes that so far only few states have complied with this obligation. M. Bourély, "Rules of International Law Governing the Commercialization of Space Activities" (1986) 30 *IISL* at 157 [hereinafter "Rules Governing Outer Space Commercialization"].
[208] Bourély holds that the only obligation of the states party to the Outer Space Treaty is to take the necessary measures in light of the commitments regarding authorization and control of the non governmental entities whose activities in Outer Space generate their responsibility. Législations spatiales nationales", *supra* note 205 at 251.
[209] Vienna Convention on the Law of Treaties signed at Vienna 23 May 1969 (entry into force: 27 January 1980) [hereafter Vienna Convention on the Law of Treaties], article 31.1.
[210] V. Kayser, "Commercial Exploitation of Space: Developing Domestic Regulation", (1992) XVII *Ann. Air & Sp. L.* at 190 [hereinafter "Developing Domestic Regulation"].
[211] *Ibid.* at 190.

almost identical language,[212] in 1963 in the UN General Assembly Resolution 1962,[213] which is considered to be reflective of customary international law. The negotiating history of Resolution 1962 and the Outer Space Treaty shows that one of the greatest concerns, aside from the militarization of outer space[214], refers to the confronted positions held by the Soviet Union and the United States with respect to the participation of private entities in outer space. The United States advocated for the absence of restrictions to the private enterprise[215], while the Soviet Union proposed an absolute state monopoly on space activities[216]. The result[217] -adopted by consensus- was a compromise between both positions[218]. Thus, the Outer Space Treaty does not forbid commercial activities but it requires that the activities of non-governmental entities be under state authorization and continuing supervision. In effect, as a *quid pro quo* for allowing the participation of private entities the Soviet Union demanded the inclusion of the obligation of states to authorize and supervise those activities[219]. This clearly shows that there was no intention on the part of the Soviet Union –let alone the United States- to oblige states to pass national law to carry out the authorization and supervision obligations. This fact by itself demonstrates that the main negotiating actors of the Outer Space Treaty did not intend to impose such

---

212 "The activities of non-governmental entities in outer space require authorization and continuing supervision by the concerned state."

213 Principles adopted by the General Assembly Declaration of Legal Principles Governing the Activities of States in the Exploration and Use of Outer Space (Adopted on December 13, 1963, Resolution1962 by the UN General Assembly 18th Session) [hereinafter "Resolution 1962"].

214 For an analysis of military aspects of space law see L. Haeck, "Le droit de la guerre spatiale" (1991) XVI Ann. Air & Sp. L. at 307.

215 M. Stuyt, *Legal Aspects of Commercial Activities of Private Enterprise in Outer Space*, (LL.M., Thesis, McGill University, 1985) [unpublished], at.4.

216 A/AC.105/C.2 16 p.2.

217 This has been considered as the primary basis for the legal admission of commercial and private endeavors in Outer Space. A. A. Cocca, Legal Framework for Economic Activity in Space (Córdoba, Advanced International Studies, 1982) at 111.

218 As explained by Dembling, "Article VI was taken almost verbatim from Article VI of the Soviet draft, which was in turn based on Paragraph 5 of the Declaration of Legal Principles. The United States draft contained no comparable provision but the United States delegate readily acceded to the Soviet version subject to changing the term "non governmental bodies corporate" to "non governmental entities, the word 'corporate' not being adequately descriptive. When the Soviet delegate accepted the minor change, debate ended on the first two sentences of Article VI." P G Dembling, "Principles governing the activities of States in the exploration and use of outer space, including the Moon and other celestial bodies", in N. Jasentuliyana & R.S.K. Lee eds., "Manual on Space Law", *supra* note 145 at 17.

219 "Commercial Space", *supra* note 7 at 51.

obligation. Neither was the intent of the other members of COPUOS.[220] In effect, there is also sufficient evidence in the COPUOS debates, as reflected in the summary records, that the form of the authorization is not prescribed in the Treaty and that therefore there is no such obligation to enact domestic legislation. As a way of example, in 1962 the representative of India held that "the control which might need to be exercised by States over such private enterprise, and the international obligations to be assumed by States in that respect, would develop out of state practice. It was likely, for instance, that most States [...] would wish to provide some system for the licensing of space vehicles.[221]"

An additional piece of evidence that the Outer Space Treaty does not require States to adopt national legislation is the practice and attitude of States vis-à-vis the prescriptions of article VI of the Outer Space Treaty. In effect, the vast majority of states have not yet enacted national legislation[222] and some spacefaring nations have only done so relatively recently[223]. The rationale behind the absence of national legislation in spacefaring countries has been that since the control can be exercised through other means there is no need for the adoption of domestic law[224]. In the case of France, for example, such control was exercised by the participation of the state itself - usually through CNES- in many activities carried out by the private sector[225]. Similar considerations can be made with respect to the former Soviet Union[226].

Authors and commentators are also of the opinion that the authorization and supervision principle does not impose the obligation to

---

[220] A/AC.105/C.2 SR.10 at 3.

[221] A/AC.105/C.2 SR.10 at 3.

[222] "Rules Governing Outer Space Commercialization", *supra* note 207 at 157.

[223] M. Gerhard, "Project 2001 Legal Framework for the Commercial Use of Outer Space," Workshop on National Space Law Legislation, Munich 5-6 December, 2000 at 1.

[224] J. Hermida, "Space Insurance: A Launch Provider's Perspective" (1997) 11 *The Air and Space Lawyer* at 1 [hereinafter "A Launch Provider's Perspective"].

[225] The CNES, a state space agency with an industrial and commercial status, created direct subsidiaries during the 1980's, such as Arianespace, Spot Image, Interspace, Novespace, Argos, Scot Conseil, etc. The authorization and supervision are exercised through active participation of CNES in the activities of its subsidiaries. P. Clerc, "French Policy and Framework" in Project 2001, *Legal Framework for Privatising Space Activities* (Cologne: Institute of Air and Space Law of the University of Cologne, 1999) at 86.

[226] Kopal, Vladimír, "The Doctrine of Space Law", in N. Jasentuliyana ed., *Space Law, Development and Scope* (New York: Praeger, 1992) at 132.

enact national space legislation. Apart from the above discussed remarks by Valérie Kayser[227], several publicists have expressed their opinion in this sense. For example, Phillip Dann recognizes that the enactment of domestic legislation is not an express requirement of the space treaties. However, he admits that "it is difficult to imagine that a party could discharge these responsibilities without introducing a system of licensing and regulations."[228]

From the foregoing discussion, it may be concluded that nothing in the *Corpus Juris Spatialis* imposes states the obligation to pass domestic legislation. The only requirement is to authorize activities in outer space and to supervise them on a continuous fashion. It may further be concluded that states are free to implement the form of such authorization and supervision, which may or may not include the adoption of national law.

3. ACTIVITIES REQUIRING AUTHORIZATION AND SUPERVISION

The first issue which we should consider is what exactly needs authorization and continuing supervision[229]. In other words, we should explore the object of the authorization and supervision principle.

Several authors have put forward that according to article VI of the Outer Space Treaty the activities that require authorization and supervision are "space activities"[230]. For the purpose of the present discussion, it is thus necessary to recall the meaning and extent of space activities. In general, they have been defined as "acts (or omissions) of individuals and entities in the exploration and exploitation process of outer space, the celestial bodies and the fruits of use of outer space"[231]. It must also be borne in mind that space activities are not, at least at present, only spatial, but rather they are planned

---

[227]"Developing Domestic Regulation", *supra* note 210 at 190.

[228] P. Dann, "The Future Role of Municipal Law in Regulating Space Related Activities", in T. L. Zwaan, ed., Space Law: Views of the Future (Deventer: Kluwer Law and Taxation Publishers, 1988) at129.

[229] It has been argued that the first issue that a national legislator must address while implementing domestic space legislation is precisely the definition of activities. "Législations spatiales nationales", *supra* note 205 at 255.

[230]A. A. Cocca, *Mantenimiento de la Utilización del Espacio Ultraterrestre con Fines Pacíficos* (Córdoba : Consejo de Estudios Internacionales Avanzados, 1986) at 5 [hereinafter "Mantenimiento de la Utilización del Espacio Ultraterrestre"].

[231] Y. M. Kolossov, "Outer Space Activities as an Object of International Space" (1985) 28 *IISL* at 233.

on Earth[232], directed from the Earth, and create results which are mostly exploited in our planet.[233]

In light of the above discussed conception of space activities, for the authors who proclaim that all space activities require authorization and supervision under the *Corpus Juris Spatialis* the authorization and supervision should encompass those activities that take place in Outer Space as well as the preparatory, related and concluding activities, which occur on Earth. Preparatory activities would include, for example, the construction of a satellite, the launch vehicle and their component parts. Related activities would include the construction and operation of ground stations and even the provision of insurance, financing, legal and consulting services. Concluding activities would cover all activities immediately following activities in outer space, such as the analysis of experiments on Earth laboratories and the refurbishing of reusable launch vehicles. Therefore, under this understanding,

---

[232] "Commercial Space", *supra* note 7 at 28.

[233] From a legal standpoint, commercial space activities can be classified as (i) accessory, (ii) principal, (iii) industrial, and (iv) complementary. P. Kahn, Philippe, "Situations d'un droit commercial spatial", in in P. Kahn, ed., *L'Exploitation commerciale de l'espace: droit positif, droit prospectif*, (Dijon: Litec Credimi, 1992) at 93; Driggers, "Space Industrialization: An overview", in B. O'Leary, ed., *Space Industrialization*, (Florida) at 2; Stuyt, *supra* note 215 at 141. The first category includes space transport. An accessory space activity is one which supports another space activity. The legal classification of an activity as accessory does not imply a negative connotation from a political, scientific or financial standpoint. The significance of space transport is clearly seen in the legal field, since many of the rules which govern this activity, its main contractual provisions and the solutions given to the problems resulting from such activity have been adopted in other spheres of Commercial Space Law. Principal activities include telecommunications, remote sensing, and data broadcasting, among others. From an economic and social viewpoint, telecommunications is certainly the activity, within this category, which has proved to be the most profitable. Income arising from remote sensing of natural resources barely exceeds ten per cent of the profits originated in the telecommunications sector. Up to the present, the only industrial activity is the manufacture of materials in outer space. Space industrialization is still in a developmental stage. The processing of materials in space includes mainly the growth of crystals and the production of pharmaceutical drugs. Both types of products have a high value per pound. It is considered that only those products with a high value per pound will be massively made in space, since in light of the costs related to a space launch, the only products which will justify such costs will be those with a high economic value and a low weight. Complementary activities are those developed only on Earth, but which are intended to aid in the performance of activities carried out in outer space. Space insurance, including satellite insurance and third party liability insurance, may be mentioned as examples.

all these activities would require authorization and continuing supervision in terms of Article VI of the Outer Space Treaty.

Nonetheless, this argument is not compatible with the Outer Space Treaty. In effect, an analysis of the language of article VI reveals that only those "activities in Outer Space"[234], i.e., only those activities which are actually carried out in outer space fall within the scope of the authorization and supervision principle. This clearly shows that the Outer Space Treaty does not mandate the authorization and supervision of all space activities but rather only those that actually take place in outer space. This does not imply that States are precluded from enacting national legislation or devising other mechanisms for the authorization and supervision of activities preceding, following or related to those that take place in outer space[235]. This, however, is not a requirement that derives from International Law, but rather it may be a response to internal political objectives of the states. In effect, as is clear from this discussion, pursuant to the prescriptions of the Outer Space Treaty, only those activities that take place in Outer Space must be authorized and supervised.

We can now move on to examine whether all of the activities in outer space must be authorized and supervised or whether some of these activities have been exempted. In this respect, it has been suggested that the national legislator must also determine which of the activities that take place in outer space require authorization and which do not[236]. Two main types of arguments have been advanced with regard to the existence of authorization and supervision exemptions to activities taking place in outer space. These arguments have been founded on technical[237] and legal grounds[238]. The former discards the need for authorization and supervision on technical grounds and the latter does it on a legal interpretation of the faculties of the appropriate state vis-à-vis the state of registry. The rationale of the technical argument appears to be that there may exist certain activities which take place in Outer Space that, because of their nature -such as some minor scientific or

---

[234] Outer Space Treaty article VI.

[235] In the Lotus case, the Permanent International Court of Justice held that "All that can be required of a State is that it should not overstep the limits which international law places upon its jurisdiction; within these limits, its title to exercise jurisdiction rests in its sovereignty" Judgment No.9 - The case of the S.S. Lotus, Collection of Judgments, PCIJ, P.19.

[236] "Législations spatiales nationales", *supra* note 205 at 255.

[237] *Ibid.* at 255.

[238] K. Tatsuzawa, "The Regulation of Commercial Space Activities by the Non-Governmental Entities in Space Law" (1988) 32 *IISL* at 83.

exploratory experiments- do not need any special measures of authorization or supervision. From a legal perspective, it has been noted that the operational activities in Outer Space are exempted from the scope of continuing supervision based upon the fact that only a State of registry retains jurisdiction and control over such activities and not the appropriate state.

It is submitted in the present study that the Outer Space Treaty does not exempt any activity which falls under the scope of Article VI from the authorization and continuous supervision requirement. While it might be argued that at the present stage of technological development the technical argument may be reasonable in some cases it is certainly contradictory of both the language and spirit of the Outer Space Treaty, which clearly requires that all activities in Outer Space be authorized and supervised at the national level. Same conclusions apply to the legal argument, for the Treaty does not foresee any exemption whatsoever.

4. CATEGORIES OF ACTIVITIES IN OUTER SPACE

In light of the above findings, the question remains as to which category of activities in outer space must a particular State authorize. In other words, it is necessary to determine the criterion for the relationship between the entity involved in an activity in Outer Space and the authorizing State.

Although the Outer Space Treaty does not specifically determine which activities in Outer Space a state may consider for authorization, publicists have concurred that this issue has to be analyzed in the context of the state responsibility provisions also contained in article VI of the Outer Space Treaty.[239] This leads to the unquestionable conclusion that states must authorize national activities, for the responsibility provisions of article VI make direct reference to such activities. In light of our earlier discussion on state responsibility, we may recall that national activities are those activities over which a state has jurisdiction, or more specifically those activities which it has the possibility to exercise legal control[240].

---

[239] Outer Space Treaty, article VI. "States Parties to the Treaty shall bear international responsibility for national activities in outer space, including the moon and other celestial bodies, whether such activities are carried on by governmental agencies or by non-governmental entities, and for assuring that national activities are carried out in conformity with the provisions set forth in the present Treaty."
[240] See *supra* note 123.

5. BASIS FOR THE AUTHORIZATION

On the basis of the foregoing discussion, it is clear that states are directed to authorize the national activities that take place in Outer Space. Neither the Outer Space Treaty nor Resolution 1962 contains any provision indicating the legal grounds for granting or rejecting the authorization[241]. Several arguments have been put forward in the literature in this respect[242]. These could be grouped in two wide categories. The first one encompasses those that advocate for the discretion of the states to decide on what legal grounds to issue the authorization[243]. The second one, which is the prevailing position, understands that it is the compliance with the provisions of the Outer Space Treaty[244], the *Corpus Juris Spatialis*[245] or even International Law in general[246].

The first arguments are centered on the fact that the enunciation of the principle in Article VI of the Outer Space Treaty does not make explicit the basis for the authorization. In this respect, it has been advanced that the states have ample discretion to authorize those activities that they consider appropriate, especially those which are compatible with their national space policy[247]. Likewise, states are entitled to refuse authorization to those activities that are irreconcilable with national space objectives and policy. As we will discuss below, this is the rationale behind the Argentine domestic space legislation[248].

The prevailing position is premised on the fact that the authorization and continuing supervision principle must be construed in a wider context of state responsibility[249]. In this respect, it must be recalled that the Outer Space Treaty prescribes that the states "shall bear international responsibility for national activities in outer space [...] whether such activities are carried on by

241 A. S. Piradov, International Space Law (Moscow: Progress Publishers, 1976) at 97.
242 See *infra* note 245.
243 A. H. Mutti, Espacio Ultraterrestre. Política y Legislación del Espacio Exterior (Buenos Aires: Ediciones Particulares, 1997) at 23.
244 *Ibid.*
245 "National Law and Commercial Activities in Outer Space" *supra* note 76 at 204.
246 "Space Treaty Revisited", *supra* note 113 at 26; M. Lachs, *The Law of Outer Space: An Experience in Contemporary Law-Making* (Sijhoff: Leiden,1972) at 22.
247 L. Cravero, *Recopilación y análisis de la normativa que regula las actividades espaciales en la Argentina* (Buenos Aires: INDAE, 2000) at 45 [hereinafter "Recopilación y análisis"]
248 See *infra* Chapter III.
249 "National Law and Commercial Activities in Outer Space", *supra* note 76 at 204.

governmental agencies or by nongovernmental entities, and for assuring that national activities are carried out in conformity with the provisions set forth in the present Treaty"[250]. Viewed in this wider context, the requirement to authorize activities is rooted in the aspiration that all activities conform to the provisions of the Outer Space Treaty[251]. In effect, to remit the legal grounds for the issuance or rejection of authorization to each state would, in practice, devoid this principle of any significant meaning.

We must still examine whether the legal grounds for the authorization of national activities in Outer Space refer to adherence to the Outer Space Treaty, the whole *Corpus Juris Spatialis* or international law in general. At first sight, the text of the Outer Space Treaty appears to be unambiguous in the sense that it requires that national activities of states conform to its provisions[252]. However, it has been argued that the Outer Space Treaty is a general agreement which enunciates comprehensive principles and that it would be inconsistent to authorize activities which comply with the Outer Space Treaty but are incompatible with other agreements[253]. It is difficult to foresee what activity could be, at the same time, permissible by the Outer Space Treaty but incompatible with another Space Law agreement. The authors sustaining this position have in mind a possible international regime to be created pursuant to article XI of the Moon Agreement[254]. Even if we

---

[250] Outer Space Treaty, Article VI.

[251] Outer Space Treaty, Article VI.

[252] Outer Space Treaty, Article VI.

[253] "Mantenimiento de la Utilización del Espacio Ultraterrestre", *supra* note 230 at 8.

[254] Moon Agreement, Article XI. Article 11 (first paragraph) of the Agreement provides that the moon and its natural resources are the common heritage of mankind, adding that the meaning of such concept must be interpreted pursuant to the provisions of the Agreement and in particular to paragraph 5 of article 11. This prescribes that states parties to the Agreement undertake to establish an international regime, including appropriate procedures, to govern the exploitation of the natural resources of the moon when such exploitation is about to become feasible. The main purposes of the international regime are established in article 11 paragraph 7. The principle of common heritage of mankind has not been defined in the Moon Agreement or in any other Space Law agreement. As shown in the US Senate hearings, some American groups expressed their fear about the fact that the expression "common heritage of mankind" had taken a particular meaning in the minds of some of those people who would presumably participate in the regime negotiations in the future. Specifically these groups argued that for too many third world countries the concept of common heritage essentially meant common property and when they negotiated the regime they would do everything to force a 'UN-style' international regime with one vote per state, under which any kind of unilateral exploitation of the Moon's resources would be forbidden on the ground that 'what

understand that such international regime may never contradict the general principles of the Outer Space Treaty,[255] we share von der Dunk's viewpoint that "with the Outer Space Treaty providing the legal framework for all space activities and laying the foundation also for further regulation, violation of its terms would amount to violation of the outer space legal regime in general[256]." Therefore, in practice, there is no distinction between requiring

---

belongs to all belongs to no one'. Another argument adopted by those who opposed the US ratification of the Moon Agreement was that the above-mentioned paragraph 5 of article 11 imposed a legal moratorium on the commercial exploitation of the Moon's resources. In effect, they argued that anything denominated as the common heritage of mankind can only be exploited with the approval of an international regime. Since the foregoing construction lacks legal grounds, for it is clear that the Moon Agreement does not forbid the development of commercial activities, other sectors voiced their opposition to the ratification of the Agreement arguing that the American private sector would not make any investments to exploit the natural resources of the Moon and other celestial bodies until the foreseen international regime was established, which would constitute in practice a *de facto* moratorium. The American Bar Association, through its International Law Section, earnestly recommended the ratification of the Agreement, subject however to certain understandings and declarations to be included in the instrument of ratification. Such declarations may be summarized as follows. First no provision in the Moon Agreement constraints the existing right of governmental or authorized nongovernmental entities to explore and use the resources of the moon and other celestial bodies, including the right to develop and exploit these resources for commercial or other purposes. In addition, nothing in the Agreement in any way diminishes or alters the right of the United States to determine how it shares the benefits derived from exploitation by or under the authority of the United States of natural resources of the moon and other celestial bodies. Second natural resources extracted, removed or actually utilized by or under the authority of a state are subject to the exclusive control of and may be considered as the property of the state or other entity responsible for their extraction, removal or utilization. Third, the meaning of the term common heritage of mankind is to be based on the provisions of the Moon Agreement, and not on the use or interpretation of that term in any other context. Recognition by the United States that the moon and its natural resources are the common heritage of mankind constitutes recognition (i) that all states have equal rights to explore and use the moon and its natural resources, and (ii) that no state or other entity has an exclusive right of ownership, property or appropriation over the moon, over any area of the surface or subsurface of the moon, or over its natural resources in place. For an analysis of this provision see, A. R. Filiato, "The Commercial Space Launch Act: America's Response to the Moon Treaty?" 10 *Fordham International Law Journal* at 778.

[255] "Commercial Space", *supra* note 7 at 86.

[256] "Fitness of International Space Law Instruments", *supra note* 115 at 4.

national activities to conform to the Outer Space Treaty or to all the International Space Law treaties and conventions[257].

The question of compliance with International Law as a whole remains pending of consideration. In this respect, Bin Cheng has put forward that in light of the fact that article III of the Outer Space Treaty prescribes that States must carry out activities in the exploration and use of outer space in accordance with International Law[258], states must assure that non governmental entities comply with international law in general[259]. This question merits the same answer given by authors that dealt with the issue of the applicability of international law to outer space activities[260]. In essence, only those norms of international law which are compatible with the principles and norms of the *Corpus Juris Spatialis* are considered to be covered by the provisions of Article III of the Outer Space Treaty. In this respect, Manfred Lachs has affirmed that:

> "this obviously implies that in all their activities in regard to and within outer space and on celestial bodies States are subject to the rule of international law. The term thus used refers to the worldwide legal system which is binding on States in all other areas of their mutual relations. None of this, however, implies an automatic extension to outer space and celestial bodies of 'international law, including the Charter of the United Nations' *in toto* [...]. Many parts of their chapters are destined for specific environment and thus do not lend themselves to application in other areas. Some rules cannot be applied to outer space *ex definitione*. Some others are of the nature of *lex specialis* for specific environments [...] The 'mechanical transfer' of institutions from one environment to another is of little avail: it may lead to distortions and even seriously stunt the development of the new branch of law".[261]

Therefore, Bin Cheng's concern[262] tends to be more theoretical than practical, for, in actuality, the legal grounds for issuing or denying

---

257 "Commercial Space", *supra* note 7 at 39.

258 Outer Space Treaty article III.

259 "Space Treaty Revisited", *supra* note 113 at 16.

260 R.S. Jakhu, *The Legal Regime of the Geostationary Orbit*, (DCL, Thesis, McGill University, 1983) [unpublished], at 163 [hereinafter "The Geostationary Orbit Regime"]; Lachs, *supra* note 246 at 15.

261 Lachs, *supra* note 246 at 21.

262 "Space Treaty Revisited", *supra note* 113 at 16.

authorization for carrying out activities in Outer Space would always eventually revolve back to the Outer Space Treaty.

In conclusion, it is submitted that the legal grounds for granting or rejecting the authorization to embark on national activities in Outer Space is the adherence –or lack of adherence, respectively- to the provisions of the Outer Space Treaty. Nevertheless, as will become apparent in our discussion of the national space norms, as a matter of strictly domestic law, States may adopt other parameters or standards for the authorization of a space activity. For example, these parameters could make reference to environmental issues, which even if they are vaguely dealt with in the Outer Space Treaty are contemplated in its spirit[263].

A further consideration which must be taken into account is that although the present discussion has limited itself to the analysis of the legal grounds for the authorization of activities in outer space, it is submitted that given the fact that the provisions governing the authorization and supervision are identical, the legal grounds for the control of those activities are also the same.

6. THE APPROPRIATE STATE

The term "appropriate State" is subject to different interpretations as the Outer Space Treaty omitted to provide a definition. The negotiating history of the treaty only shows that its text was taken from the Soviet draft, which in turn had been inspired in Resolution 1962, and accepted without substantive debate[264]. The final language of the treaty contained the same text as these two instruments, except that it referred to the "state concerned" instead of the "appropriate state"[265].

In the specialized literature there are divergent opinions and perspectives around the construction of this concept. In this respect, the appropriate state has been equated with the state of nationality[266], the launching state[267], the state concerned[268], the responsible state[269] and with a combination of

[263] Mantenimiento de la Utilización del Espacio Ultraterrestre", *supra* note 230 at 5.
[264] B. Cheng, "Convention on International Liability for Damage Caused by Space Objects", in "Manual on Space Law", *supra* note 145 at 101.
[265] *Ibid.* at 101.
[266] "Liability Overview", *supra* note 131 at 377.
[267] "Problems of State Responsibility", *supra* note 115 at 140.
[268] "Space Treaty Revisited", *supra* note 113 at 14.

these concepts[270]. The doctrines of essential role and substantial participation have also been formulated to explain this concept. Additionally, there have been many discussions in the literature on whether there should be one or several appropriate states[271]. With respect to this latter issue, there seems to be a general consensus that even if the text of the Outer Space Treaty refers to this term in the singular, there may be several "appropriate States" with respect to a particular activity in outer space.[272] In effect, since there are a number of States which may be potentially responsible for a certain space activity, it is understood that there must necessarily be several appropriate States. Otherwise these States could be held responsible for an activity which they were deprived to exercise the authorization and supervision functions[273].

As concerns the scope of the notion of appropriate state, Gorove understands that the appropriate state is the state of nationality. He reasons that article VI of the Outer Space Treaty places responsibility on states for their national activities and since the authorization and the responsibility principles are strongly interconnected the appropriate state must be construed as the state of nationality[274]. Nonetheless, Gorove himself realizes the flaw of this argument since he acknowledges that had the drafters intended to equate the appropriate state with the state of nationality they could have simply used the expression state of nationality[275]. However, it is beyond doubt that both the authorization and the nationality principle are interrelated and that in many instances the State which is responsible for the activities of a non governmental entity will also be the appropriate state.

Van Traa Engelman holds that the appropriate state should be assimilated with the concept of launching state. The underlying rationale behind this premise is that this is compatible with the practice of effective control by a state over its own territory[276]. For van Traa Engelman this would cover nearly all

---

269 Kerrest, "Remarks on the Responsibility and Liability for Damage Caused by Private Activity in Outer Space" (1998) 40 *IISL* at 139 [hereinafter "1998 Remarks"].

270 M. Bourély, "Rules of International Law governing the Commercialization of Space Activities"(1985) 29 *IISL* at 159.

271 "Space Treaty Revisited", *supra* note 113 at 14.

272 For details, see Bockstiegel, K.-H., "The Term 'Appropriate State' in International Space Law", (1994) 37 *IISL* at 79 [hereinafter "The Term Appropriate State"] and W. B. Wirin, "Practical Implications of Launching State - Appropriate State Definitions", (1994) 37 *IISL* at 109.

273 B. Cheng, *Studies in International Space Law* (Oxford: Oxford University Press, 1997) at 609.

274 "Liability Overview", *supra* note 131 at 377.

275 *Ibid.* at 377.

276 "Problems of State Responsibility", *supra* note 115 at 141.

cases of space activities, excluding only those instances in which private organizations launch space objects from other places than a State's territory or facility[277]. To fill the lacunae represented by these cases, van Traa Engelman proposes the application of a second criterion based on the nationality of the persons or the seat of the organization that launches the space object[278].

Bin Cheng holds that since for each activity in Outer Space there may be several States which are concerned -especially because they may bear international responsibility- it is legitimate for all these States to subject those activities to their authorization and continuing supervision. In this respect, Cheng understands that all those States may be considered the appropriate state[279]. He recognizes, however, that the terminology used in Resolution 1962 was more adequate than the one adopted in the Outer Space Treaty[280].

Herczeg's conception on the scope of the term appropriate state is based on an analogy of his understanding of the meaning of launching state[281]. For Herczeg the distinctive criterion for the determination of the launching state is the standard of essential role. Thus, a state may be considered launching state when it plays an essential role in a space activity. Under this line of reasoning, the four criteria contained in article VII of the Outer Space Treaty appear to be mere examples of the essential role theory with regard to space endeavors. Extrapolating this conception to the notion of appropriate state, Herczeg concludes that several states may have an essential role with respect to each activity. As examples of these states, he includes the state of the seat of non governmental entities, the state of jurisdiction and the launching state[282]. Herczeg's position has been shared by Barton, who also understands that "the appropriate state could be any state which played an important role in the launching of the spacecraft[283]."

---

[277] "The launching state criterion would remove much of the obscurity surrounding article VI and would offer an effective means of realizing the aims of the State responsibility principle." *Ibid.* at 141.

[278] *Ibid.* at 141.

[279] Bin Cheng is of the opinion that "every State Party concerned is an appropriate State, and while it is possible, when more than one State is involved, to arrange for one of them to carry out this function, it remains the responsibility of every contracting Party involved to see that it is in fact carried out by an appropriate State Party. "Space Treaty Revisited", *supra* note 113 at 14.

[280] *Ibid.* at 28.

[281] I. Herczeg, "Problems of Interpretation of the Space Treaty of 27 January 1967", (1967) 10 *IISL* at 107.

[282] *Ibid.* at 107.

[283] Barton, "Summary of Discussions", (1967) 10 *IISL* at 116.

Wirin also elaborates on the essential role standard, but he refers to it as substantial participation and understands that the concept of appropriate state requires substantial participation on the part of the states. Thus, states must have a controlling role in the decisions concerning activities in Outer Space in order to qualify as appropriate states. In this respect, he argues that those states that are "merely used for convenience" may not be considered as appropriate States and suggests that States that manage, control or are the principal investors in a space project must be deemed appropriate states[284].

For Kerrest, the concept of responsible state is identical to the concept of appropriate state. In light of the interplay between the concepts of responsible state and appropriate state, it has been held that a State must authorize and supervise the activities over which it is responsible[285].

In the understanding that the term appropriate state allows for several interpretations due to the vagueness of its formulation, Bourély exemplifies his position by enumerating the States which could qualify as appropriate states in terms of article VI of the Outer Space Treaty[286]. In this respect, he points out that the state which exercises jurisdiction and control of a non governmental entity carrying out a space activity and the state that governs the territory in which that entity has its head office, establishment, branch or plant could be appropriate states. He also considers as appropriate states the state from which territory a space vehicle is launched or a satellite is controlled or operated. Other examples of appropriate states are the launching state, the state of registry and the state which owns the space device[287].

All these examples permit to conclude that there are many situations where a state may qualify as appropriate state. Böckstiegel is of the opinion that all the arguments exposed in the literature may have some merit and that no argument is convincing enough to disregard all the other ones as acceptable.[288]

From the foregoing discussion, it is abundantly clear that the notion of appropriate state is open to several interpretations. Nevertheless, certain guidelines and conclusions may be drawn. First, there may be several "appropriate States" with respect to each particular activity in outer space.

---

[284] Wirin, *supra* note 272 at 113.

[285] Kerrest, "Remarks on the Responsibility and Liability for Damage Caused by Private Activity in Outer Space" (1997) 42 *IISL* at 139.

[286] M. Bourély, "Rules of International Law Governing the Commercialization of Space Activities"(1985) 29 *IISL* at 159.

[287] *Ibid.* at 159.

[288] "The Term Appropriate State", *supra* note 272 at 79.

Second, the common feature underlying this discussion is that many or all of these interpretations may be valid according to each circumstance. Thus, it is submitted that the determination of the appropriate state or states must be done on a case by case basis in light of the specific characteristics surrounding each activity that takes place in outer space. In summary, therefore, the recommended approach to be undertaken to determine the appropriate state is to look at the particular activity and to identify all those States that are concerned with such activity on account of the reasons and criteria expounded above.

7. SUBJECT OF THE OBLIGATION TO AUTHORIZE AND SUPERVISE

It is now necessary to inquire who should carry out the obligations to authorize and supervise the national activities in outer space. The Outer Space Treaty simply proclaims that it is the appropriate state[289] that should do so. However, it does not give any further guidelines as to the actual entity within a state that must actually grant the authorizations and exercise the supervision.

State practice in the space arena has shown that in general the authorization and supervision functions have been assumed by the Executive Branch[290] or by an agency or another subdivision of the Executive Branch[291]. Nevertheless, given the different solutions and the low number of states that have adopted formal mechanisms to carry out the authorization and supervision functions it is submitted that this practice is not sufficient to constitute an international custom[292]. What may, nonetheless, be gathered from this practice is that states consider that the question of who must grant

---

[289] From a legal standpoint a state is considered as a subject of international law which possesses a permanent population, a defined territory, a government and capacity to enter into relations with other states. Convention on Rights and Duties of States, signed 26 December 1933. Thus, state is a general and abstract concept - characterized as an organization employing specialized personnel, which controls a consolidated territory and is recognized as autonomous and integral by the agents of other states.

[290] See *infra* Chapter II.

[291] See *infra* Chapter II.

[292] Article 38 of the United Nations Charter refers to international custom as evidence of a general practice accepted as law. United Nations Charter, article 38. According to Brownlie, it is a general recognition among states of a certain practice as obligatory. Brownlie, *supra note* 104 at 5.

the authorization is an issue which is left to the discretion of the states[293]. This conclusion is in line with our general understanding that states are free to implement the form of such authorization and supervision[294].

While the foregoing has not created much controversy, the possibility of delegation has originated some concern among specialized authors, particularly during the Cold War years. For example, Piradov and other Soviet commentators have expressed their objection to the possibility of a state's delegating authority to carry out the obligations of permission and supervision[295]. In a similar vein, Jenks understands that a general delegation of authority in space matters to a private body would clearly be incompatible with these obligations of the states[296]. Jenks, however, does not explicate the reasons why this would be so[297].

We are of the view that the question of the possibility of delegation is also subsumed in our foregoing conclusion regarding the states' freedom to adopt the manner of the authorization and continuing supervision of activities in outer space[298]. Therefore, the delegation of authority to any entity of the state's choice would not contradict the principle consecrated in article VI of the Outer Space Treaty. The state would, however, retain responsibility for the actual authorizations given by the delegated entity and for the supervision it carries out as if the state itself performed these functions. In line with this conclusion, a state could, for example, delegate authorization and supervision to an independent entity, such as the organs which are entrusted to oversee telecommunications activities in many jurisdictions or even to non governmental organizations.

8. CONTINUING SUPERVISION

The scope and nature of the supervision have not been clarified in the Outer Space Treaty, except for the fact that the Treaty requires that the supervision be continuing. An examination of the object and purpose of the supervision as discussed in the literature will help us elucidate this issue.

---

293 In this respect, the authorization could, for example, be assigned to the Legislature and the supervision to the courts which could act in case of a charge or request filed by an interested party or the society as whole, such as in the case of environmental problems.
294 See *supra* note 210.
295 Piradov, *supra* note 241 at 97.
296 C. W. Jenks, Space Law (London: Stevens, 1965) at 211.
297 *Ibid.* at 211.
298 See *supra* note 227.

According to Hosenball, the supervision required by article VI should be conducted so as to permit "the supervising state to effectively control" the activities of the non governmental entities. [299] This implies that a government representative does not need to be continuously present at the premises of the non governmental entities[300], which in many jurisdictions would be incompatible with constitutional guarantees[301]. For Menter, this should be done through the issuance of regulatory directives by a state agency within legislative enacted guidelines and by consultations, reports, inspections and by investigation of reported discrepancies. He further suggests that non compliance should be sanctioned[302].

In light of our conclusions regarding the freedom of states to adopt the form of the authorization and supervision, it is submitted that this freedom also extends to the frequency and other modalities of the supervision, provided, however, that it is carried out in a continuing fashion[303]. In this respect, Menter's inventory could serve as a basis for structuring the supervision. States are, nonetheless, free to select other measures. In this same vein, States are also free to interpret the meaning of the term "continuing", so long as they do not denaturalize the general meaning given by the drafters of the Outer Space Treaty to this concept[304].

9. NON GOVERNMENTAL ENTITIES

Like with other concepts of article VI, the Outer Space Treaty does not provide a definition of the phrase non governmental entities. Its meaning may not be found elsewhere in the Treaty or in the records of the negotiating history[305]. It is clear, however, that the intention of the Soviet delegation, which negotiated this text with the United States, was that states assumed international responsibility for

---

[299] S. N. Hosenball, "The Law Applicable to the Use of Space for Commercial Activities", (1983) 26 *IISL* at 144.

[300] M. Menter, "Legal Responsibility for Outer Space Activities", (1983) 26 *IISL* at 122.

[301] Argentine Constitution article 14 freedom of commerce.

[302] Menter, *supra* note 300 at 122.

[303] Based on our previous findings, the object of the supervision, i.e., what states have to supervise, is already answered. It has been submitted that states have to control whether the activities of the non governmental entities adhere to the provisions of the Outer Space Treaty.

[304] In France the activities carried out by the CNES are controlled on a posteriori basis and not on a priori fashion. Clerc, *supra* note 225 at 84.

[305] P G Dembling, "Principles governing the activities of States in the exploration and use of outer space, including the Moon and other celestial bodies", in "Manual on Space Law", *supra* note 145 at 17.

the activities of their private national companies[306]. However, there are at present a myriad of entities which at first sight do not qualify as entirely non governmental or governmental. For example, many types of entities have engaged in outer space activities in the last decades. These include mixed companies[307], i.e., companies which are held by private shareholders and the government, research institutes inserted within state-owned universities[308], commercial companies whose sole shareholder is the state or a state agency[309], and international consortia made up of states, private companies and mixed entities[310].

The purpose of the provision under analysis is that space activities be developed on the level of relations between States and not on the level between States and private companies[311]. The Soviet Union's concern, as reflected in the drafts presented, was that private entities engaged in the commercial space activities without a control of the States[312]. Thus, it is submitted that as long as the State –acting as such or through an agency or any other legal structure- is substantially engaged in a space activity it is not obliged to expressly authorize and supervise that activity since its sole participation may be considered an authorization and a permanent supervision. This conclusion is supported by the practice of France, one of the main spacefaring nations[313].

10. MOMENT OF AUTHORIZATION

The Outer Space Treaty is also silent with respect to the moment when the authorization must be given. However, the analysis of the negotiating history of an analog situation serves to clarify the intent of the Outer Space Treaty drafters.

A similar issue arose with the adoption of the Registration Convention, which helps to interpret this question. This Convention prescribes that when a space object is launched into outer space the launching state has to register the space object by means of an entry in an appropriate

---

[306] Piradov, *supra* note 241 at 97.
[307] Gerhard, *supra* note 223 at 1.
[308] Asociación Argentina de Tecnología Espacial.
[309] New Zealand Airways Corporation.
[310] "Commercial Space", *supra* note 7 at 26.
[311] Piradov, *supra* note 241 at 97.
[312] P G Dembling, "Principles governing the activities of States in the exploration and use of outer space, including the Moon and other celestial bodies", in "Manual on Space Law", *supra* note 145 at 17.
[313] Clerc, *supra* note 225 at 84.

registry which it has to maintain. However, it did not make specific reference to the time in which the information is to be given, i.e., before or after the launch. In this respect, it has been put forward that States are free to decide on the time to make the notification to the Secretary-General of the United Nations. On the other hand, some publicists have argued that in light of the object and the purpose of the Convention, the notification, if possible, should be made prior to the launching[314]. In the case of the principle of authorization and continuing supervision its purpose is even clearer. In effect, since the international legislator sought that the States ensured that the activities of non governmental entities conform to the norms of the Outer Space Treaty, this purpose may not be achieved except if States secure compliance with the norms of the Treaty before these activities take place. This, however, does not imply that the Outer Space Treaty imposes States the obligation to authorize and supervise preparatory or related activities. As discussed above, only activities that occur in outer space require authorization and supervision. However, on account of the purpose of this principle the authorization may only be granted or refused before the commencement of said activities.

11. RELATIONSHIP WITH THE OUTER SPACE TREATY

The above discussions have intended to shed light on the concepts contained in the formulation of the authorization and continuing supervision principle. We will now proceed to analyze other relevant principles of International Space Law in order to adequately contextualize this principle against the mosaic of principles contained in the Outer Space Treaty. This examination is of significant importance for States since several principles contained in the *Corpus Juris Spatialis*, as well as in International Law, may affect the content of national legislation. Since States must ensure that the activities of their nationals adhere to the provisions of the Outer Space Treaty, we will analyze which of these principles a State has to take into account –an up to what extent- when considering the issuance of the authorization to its national non governmental entities.

12. FREEDOM OF EXPLORATION AND USE

Article II second paragraph of the Outer Space Treaty contains the freedom of exploration and use principle, which reads as follows "Outer space, including the moon and other celestial bodies, shall be free for exploration and use by all States without discrimination of any kind, on a basis of equality and in

---

[314]A. A. Cocca, "Registration of Space Objects", in "Manual on Space Law", *supra* note 145 at 180.

accordance with international law, and there shall be free access to all areas of celestial bodies."[315] This principle is of fundamental importance as it sets the rule of freedom in space for exploration and use[316]. It declares outer space open for innumerable possibilities and a myriad of activities[317]. By means of the adoption of this principle, States have an affirmative obligation to permit other States to embark on space activities and to remove all barriers which may obstruct the smooth development of activities in Outer Space by other States. Its enunciation has been carefully formulated in broad terms so as to embrace the widest range of possible activities[318]. It is intended as a statement of goal and purpose so as not to subject space to exclusive appropriation by any particular power[319] and to foster the exploration and use of outer space[320].

This principle has been considered a fundamental part of the *Corpus Juris Spatialis*[321] and one of the pillars on which the whole system is based upon[322]. Even before the beginning of the space age, authors recognized that any regulatory environment for outer space activities should be founded on the principle of free use[323]. This principle was first recognized by the UN General Assembly in 1961 in its Resolution 1721.[324] It was then incorporated in Resolution 1962[325] and finally included in the Outer Space Treaty[326].

---

315 Outer Space Treaty, article II.

316 US Senate, Hearings Before the Committee on Foreign Relations, 90th Congress, First Session, March 7, 13 and April 12, 1967 (Washington: 1967 US Government Printing Office) at 53 [hereinafter "US Senate Hearings 1967"].

317 According to Cocca, the term "free" is not used in the political sense, such as national freedom, but rather it refers to the freedom to make use of a natural phenomenon within the universe. A. A. Cocca, "Principles for a Declaration with Reference to the Legal Nature of the Moon", (1957) 1 *IISL* at 14.

318 "US Senate Hearings 1967", *supra* note 316 at 53.

319 *Ibid.* at 10.

320 "Commercial Space", *supra* note 7 at 39.

321 Ambrosini considered at COPUOS that this principle is not only fundamental but also historic, for it marked the first time an international issue of such importance was adopted by all states. A/AC.105/C.2/SR.2 p.2, "The Geostationary Orbit Regime", *supra* note 260 at 23.

322 M. A. Ferrer, *Derecho Espacial* (Buenos Aires: Plus Ultra, 1979) at 55 [hereinafter "Ferrer Derecho Espacial"].

323 Schacter, Who Owns the Universe? Space Law Symposium, Special Committee on Space and Astronautics, US Senate, 85th Congress, 2nd Session, 1957, 8-7 quoted by N. M. Matte, Space Activities and Emerging International Law (ed.) (Montreal: McGill University, 1984) at 249.

324 UNGA Resolution 1721 (XVI). "Outer Space and celestial bodies are free for exploration and use by all States in conformity with international law."

325 Resolution 1962. Piradov, *supra* note 241 at 83.

13. EXTENT AND SCOPE OF THE FREE EXPLORATION AND USE PRINCIPLE

The freedom of exploration and use principle[327] encompasses two positive aspects: (i) the right of free exploration[328]; and (ii) the right of free use[329]. It has been debated whether this principle also includes the right to free exploitation[330], which is a concept that denotes commercial utilization of outer space[331]. While the view of the publicists has been divided[332], an examination

---

326 Jakhu explains that "at the time of the Outer Space Treaty was negotiated, the freedom principle did not pose any serious problems of acceptance." Jakhu mentions the following reasons. First, the International Geophysical Year resulted in the most intensive observation of the Earth. Second, States allowed the free passage of space objects over their territories. Third, it had already been declared in Resolution 1962. "The Geostationary Orbit Regime", *supra* note 260 at 146.

327 Outer Space Treaty, article 1.

328 *Ibid.*

329 N. M. Matte, *Space Activities and Emerging International Law* (ed.) (Montreal: McGill University, 1984) at 270. It has been argued that the freedom of exploration and use also encompasses the freedom of information which is elaborated on article XI of the Outer Space Treaty. The Treaty mandates states to inform the Secretary General of the United Nations as well as the public and the international scientific community, to the greatest extent feasible and practicable, of the nature, conduct, locations and results of space activities. Non governmental entities have a right to have access to this information and the obligation to provide information regarding their space activities to the state which has jurisdiction on them, which in turn has to inform the Secretary General and the general community. This principle has been one of the objectives of the US diplomacy which advocated for a broad exchange of information. However, it faced intense objection from the outset on the part of the Soviet Union, the result being a compromise that information should be shared to the extent feasible and practical on only certain aspects of space endeavors. Committee on Foreign Relations, United States Senate, Treaty on Outer Space, 90th Congress, First Session, March 7, 13 and April 12, 1967, (Washington: U.S. Government Printing Office, 1967) at 42.

330 G. Marcoff, *Traité de Droit international public de l'espace* (Fribourg, 1973) at 330.

331 The enunciation of the principle has also raised other discussions. For example, it has been debated whether the right of free exploration has a wider scope than the right to free use. Matte argues that the concept of exploration is associated with scientific investigation and use is more closely connected with exploitation and application. Thus, since paragraph 3 of article I enunciates the principle of freedom of scientific investigation without the limitations contained in paragraph 2 of that article, scientific activities, i.e., exploratory, enjoy a preferential status over the application activities. N. M. Matte, Space Activities and Emerging International Law (ed.) (Montreal: McGill University, 1984) at 271.

of the negotiating history and the debates in COPUOS clearly indicates that when this issue was debated the term use was meant to include the use and exploitation of outer space and its natural resources, as is clear from the interpretation of the French representative in COPUOS, who unambiguously explained that the term "use" should also cover "exploitation"[333].

It has also been put forward that the components of the right of free use include: (i) the freedom to choose the physical location of an exploitation, (ii) the freedom to choose the methods of use, and (iii) the freedom to choose the method of managing the activity.[334] While these components are undoubtedly included within the free use principle[335], it is submitted that in light of the characteristics of the freedom of exploration and use principle examined above[336], the scope of the concept of free use is more profound and extends to every conceivable aspect of a space activity. In line with this argument, it is considered that the selection of the launch vehicle for satellite operators and the election of launching facilities for launch service providers, the administration of risks –to the extent permitted by law or other international or regional agreements[337]-, the election of personnel, the method of financing[338], and the legal structure of the endeavor are also included, among many other aspects, within the scope of the free use principle.

The Outer Space Treaty does not specify what activities are covered by the freedom principle[339]. It simply states that outer space will be free for exploration and use[340]. In the discussion on the extent of the activities covered by the authorization requirement contained in article VI it was concluded that only activities that took place in outer space are included and that consequently preparatory, related and concluding activities do not require authorization and supervision[341]. It could be held that the freedom of exploration and use principle covers the same activities. However, that interpretation would be incompatible with the purpose of the Treaty. In effect,

---

[332] Marcoff, *supra* note 350 at 330. For a different position, "The Geostationary Orbit Regime", *supra* note 260 at 151.

[333]Christol, *supra* note 153 at 39.

[334]Nordlund, Frédéric, *Le régime juridiques des activités industrielles et commerciales conduites dans l'espace extra-atmosphérique: nouvelles orientations* (LLM, Thesis, McGill University, 1990) [unpublished] at 69.

[335] Outer Space Treaty, article 1.

[336] See *supra* note 321.

[337] "Space Risk Management" *supra* note 60.

[338] J. Hermida, "Space Financing", (1998) 13 *The Air and Space Lawyer* at 1.

[339] Outer Space Treaty, article 1.

[340] *Ibid.*

[341] See *supra* note 234.

the purpose of the treaty is to permit the widest range of possible activities in the exploration and use of outer space.[342] Therefore, it must be concluded that the freedom principle necessarily embraces those preparatory, related and concluding activities, for otherwise, i.e., if a State were prohibited, for example, to carry out preparatory activities it would make the freedom principle absolutely illusory as it would be impeding their nationals to enjoy the freedoms granted by the Outer Space Treaty. In the same vein, all types of activities, e.g., remote sensing, satellite telecommunications and space transportation, among others, are covered and protected by the freedom principle.

Since the freedom and exploration principle benefits only the States and not their private firms, in those countries with a liberal market economy and where their space policy pursues the development of the private sector space industry, it would be desirable that these States extend –to the maximum possible extent- the guarantees of the freedom principle to their private firms. In such case, wide access to outer space would undoubtedly foster the growth of the private industry.

14. LIMITS TO THE FREEDOM PRINCIPLE

The freedom principle, like any other right[343], is not absolute and is subject to several limits[344] expressly contained in the Outer Space Treaty[345] and other constraints derived from general international law[346]. It is essential to examine the restrictions to the freedom of exploration and use principle. This is of substantial practical importance since it will indicate the barriers that States may impose on non governmental entities.

Since the analysis of these principles has been the object of extensive writings,[347] the present study will very briefly discuss only their main features in connection with their implication to the authorization and supervision of activities at the national level.

---

[342] "US Senate Hearings 1967", *supra* note 316 at 10.

[343] B. Cardozo, "Mr. Justice Holmes", (1931) 44 *HARV. L. REV.* at 682.

[344] According to Lauterpacht, "There is no legal right, however well established, which could not, in some circumstances, be refused recognition". Lauterpacht, International Law, quoted by I. Brownlie, Brownlie, *supra note* 104 at 432.

[345] See *infra* note 363.

[346] See *infra* note 390.

[347] "ILA Report", *supra* note 82.

15. OTHER PRINCIPLES OF INTERNATIONAL SPACE LAW

15.1. Benefit of mankind

It has been argued that the benefit of mankind doctrine[348] constitutes the purpose of the freedom principle and that for this reason the freedom principle may not be used as a justification for arbitrary or illegal activities.[349] In line with this argument, the freedom to explore outer space is subject to the condition that the activities be carried on in the benefit and interest of mankind. This means that those activities, whether governmental or not, which are not beneficial to mankind may not be carried out[350].

However, since the benefit of mankind principle has not been defined in the *Corpus Juris Spatialis* it does not give a clear guideline about the required type of benefit for a space activity to be in conformity with this principle[351]. It is thus submitted that since this principle has been enunciated as only a general goal[352] space activities are in compliance with this principle as long as they contribute in a general sense to the social welfare.[353] In practice, it is considered that the common benefit doctrine may not impose any serious limitation to the vast majority of space activities[354].

15.2. Non-appropriation

Article II of the Outer Space Treaty embodies the non-appropriation principle, which establishes that "outer space, including the moon and other celestial bodies, is not subject to national appropriation by claim of sovereignty, by means of use or occupation, or by any other means."[355] This principle clearly

---

[348] A. A. Cocca, "Common Heritage of Mankind: A Basic Principle of the International Legal System", (1988) 31 *IISL* at 89 [hereinafter "CHM"].

[349] G. P. Zhukov & Kolossov, *International Space Law* (New York: Praeger, 1984) at 42.

[350] Thus, for example, a mission which intends to develop heavy weapons in outer space will not be legal. This is so not because it will contradict the principle of peaceful use of outer space, but because it does not offer any benefits to mankind.

[351] "CHM", *supra* note 349 at 89.

[352] "US Senate Hearings 1967", *supra* note 316 at 10.

[353] This is so even if their main purpose is the mere obtainment of profits. Advocates of a wide and comprehensive construction of this common benefit doctrine have also argued that this entails a limitation for commercial space activities. Tatsuzawa, *supra* note 238 at 343.

[354] "Commercial Space", *supra* note 7 at 43.

[355] Outer Space Treaty, article II.

sets a limitation to the freedom of exploration and use. Therefore, at the national level a state may not authorize activities which seek the national appropriation of outer space in terms of article II of the Treaty. Thus, it is our view that this principle precludes the possibility of appropriation of Outer Space and celestial bodies by means of private property[356].

However, the principle of non-appropriation is not absolute and it does not imply the disregard of certain rights on some areas of outer space, e.g., the right to use a specific orbital position (as long as the rules of International Law are observed)[357], the right to use a specific area where a space station is built[358], or the space vehicle's right to its trajectory[359], among others. The legitimate exercise of these rights of use implies the recognition of a sort of *de facto* ownership, which does not seem to contradict the true spirit of the non-appropriation principle[360].

15. 3. Peaceful Activities

The Outer Space Treaty prohibits States to place in orbit around the earth any objects carrying nuclear weapons or any other kinds of weapons of mass destruction, and to install these weapons on celestial bodies, or station such

---

[356]This fact does not imply the non-existence of private property in Outer Space. On the contrary, in accordance with article VIII of the Space Treaty, the ownership of space objects, even those built in outer space, does not change while such objects are in outer space. Thus, for example, a launched vehicle owned by a private space launch carrier launched into outer space pursuant to the international provisions in force will still be owned by that carrier and that carrier's rights will be recognized by all the states and non-governmental entities.

[357]"The Geostationary Orbit Regime", *supra* note 260 at 151.

[358] A. A. Cocca, "Preface", in J. Hermida, *Commercial Space Law: International, National and Contractual Aspects* (Buenos Aires: Ediciones Depalma, 1997).

[359] The right to the trajectory is a concept created by the Argentine Space Law school, which applies to any space vehicle. Ferrer held that "if a spacecraft registered in the National Registry set forth by article II of the Registration Convention, duly informed to the Secretary General of the United Nations pursuant to the provisions of article IV of said convention, flies through outer space in the orbit which has been informed to the United Nations Registry, in accordance with the requirements of article IV first paragraph item d) of such Convention, creates an *erga omnes* obligation to respect its trajectory, which impedes any interference." Ferrer finds the foundations of the right to the trajectory in the principles of International Space Law, particularly in the second part of article I of the Space Treaty, which establishes that Outer Space will be free for exploration and use. M. A. Ferrer(h)., "El derecho a la trayectoria", (1997) 13 *IISL* at 160.

[360] Peyrefitte, *supra* note 64 at 50.

weapons in outer space in any other manner[361]. Additionally, it prescribes that the moon and other celestial bodies must be used exclusively for peaceful purposes, where the establishment of military bases, installations and fortifications, the testing of any type of weapons and the conduct of military maneuvers on celestial bodies are strictly forbidden[362]. Therefore, States may not authorize national activities which conflict with the peaceful use of outer space principle as formulated in the Treaty[363].

15. 4. International cooperation

It has been argued that the principle of international cooperation also entails a restriction to the freedom of exploration and use of outer space[364]. The traditional Argentine school of Space Law[365] has considered that the principle of international cooperation[366] has a compulsory nature[367]. For traditional

---

[361] Outer Space Treaty, article 4.

[362] Outer Space Treaty, article 4. Thus, the Treaty does not require that activities carried out elsewhere in outer space be exclusively peaceful. The Space Treaty merely states that the activities must be carried on pursuant to international law and in the interest of maintaining international peace and security.

[363] R. A. Ramey, "Armed Conflict on the Final Frontier: The Law of War in Space" (2000) 48 *A.F. L. Rev.* at 1.

[364] "Ferrer Derecho Espacial", *supra* note 322 at 86.

[365] A. A. Cocca, Recent Developments in Space Law and Space Policy in Latin America, (1996) 16 *ECSL News.*

[366] International cooperation in the cosmos exploration has been defined by Elena Kamenetskaia as the combined activity of states in the scientific-technical, economic, political and legal spheres, to ensure the study and use of the outer space and the accomplishments of cosmic science and technology for peaceful purpose, in benefit of all the states and peoples. As inferred from the above conceptualization, international cooperation in space involves different dimensions: political, legal, economic and technical. Cooperation in political matters tends to create the international favorable conditions which ensure the exploration and use of outer space for peaceful purposes and to favorably solve the international legal problems of exploration and exploitation of space. State cooperation in the legal field facilitates the elaboration and the validation of the political-legal principles which determine the outer space system, the objectives of space activities, as well as the reinforcement of the rule of law in outer space and the development of International Space Law. Economic cooperation originates from the high costs implied in every space activity and in the high risks for the investments in space endeavors. The purpose of technical cooperation is to gather and pool the efforts of states in the implementation of scientific and technical research in the use of outer space, and in the development of space activities, including, launch, telecommunications, and manufacturing, among many others. E. Kamenetskaia, "Cosmos. Cooperación. Derecho" *Ciencias Sociales*, (1983) 1 Academia de Ciencias de la URSS at163.

Argentine scholars, this principle creates a legal obligation on the part of the participants to an activity in outer space. Thus, in order for an activity to be legal it must be the result of international cooperation.[368] According to Cocca, international cooperation pursuant to the Outer Space Treaty and to all the subsequent international instruments, is a legal obligation, which conditions the lawfulness of every space activity[369].

This doctrine has faced a strong opposition since it does not have any legal grounds. As put forward by Miklódy[370] "in most cases, international cooperation is simply *an obligatio de contrahendo* and not an unconditional duty". In this regard, Kolossov[371] holds that no State shall impose upon another state the subject and the terms of cooperation in one or another area. Cooperation is rather the result of bilateral and multilateral agreements at the State level and the result of a corporate decision in the private sector. Therefore, it is submitted that notwithstanding the importance of international cooperation, it only constitutes a necessity and a beneficial advantage in outer space activities, but it may not be deemed as an obligation whose non compliance triggers off the illegality of the activity. Accordingly, it does not constitute a positive obligation and States may not refuse authorization of a national activity on the grounds that it does not seek international cooperation or that it is not the result of an international cooperative endeavor.

15. 5. Avoidance of harmful contamination

According to article IX, all space activities have to be conducted so as to avoid harmful contamination and also adverse changes in the environment of the Earth resulting from the introduction of extraterrestrial matter[372]. Provisions contained in this principle are rather vague. For example, reference to harmful contamination may appear to suggest that non harmful

---

367 A. A. Cocca, "Preface", in J. Hermida, *Commercial Space Law: International, National and Contractual Aspects* (Buenos Aires: Ediciones Depalma, 1997).

368 "Ferrer Derecho Espacial", *supra* note 322 at 77.

369 A. A. Cocca, "El espacio ultraterrestre. Labor de las Naciones Unidas en la codificación del espacio", in Problemas contemporáneos de la actividad aeronáutica y del espacio, Infante, M.T. and Irigoin, Jeannette, (eds.), *Colección de Estudios Internacionales* (Santiago: Editorial Universitaria, 1978) at 146.

370 M. Miklódy, "International Cooperation. A Legal Obligation in the Law of Outer Space?", (1983) 26 *IISL* at 231.

371 Y. Kolossov, "Cooperación Internacional en los Medios de Comunicación", (1978) 2 *Estado y Derecho Soviético* at 32.

372 Outer Space Treaty, article XI.

contamination is allowed[373]. Similarly, reference to the phrase adverse changes is not altogether clear. Furthermore, this principle refers only to harmful contamination of the Earth. It thus seems to permit contamination of Outer Space[374].

Nonetheless, this principle imposes the obligation to the States parties to the Treaty to assure that their non governmental entities comply with the non contamination provisions of the Treaty. Therefore, States are expected to consider the environmental aspects for the authorization and supervision of national activities in outer space.

General International Environmental Law also imposes restrictions to space activities, particularly in the event of harmful contamination of the Earth. This also imposes limitations to outer space activities which domestic space law must take into account[375].

15. 6. Non interference

Article IX of the Outer Space Treaty prescribes that in the exploration and use of outer space, States must have due regard to the corresponding interests of all other States[376]. Furthermore, in the event that a State has reasons to believe that an activity or experiment planned by another State or its national in outer space would cause potentially harmful interference with its space activities it may request consultation concerning the activity or experiment[377]. The obligation not to interfere with the activities of other States and to have due regard for their interests constitutes another factor, which must be adequately considered by States at their domestic level.

16. ABUSE OF RIGHTS

The abuse of rights principle prescribes that States may not exercise a right in order to cause damage to another state or in a way which may impair the rights of other states[378]. Put differently, the right of freedom of use of outer

---

373 S. M. Williams, *Riesgo ambiental y su regulación* (Buenos Aires: Abeledo-Perrot, 1998) at 55.
374 *Ibid.* at 56.
375 M. Gerhard & K.U. Schrogl, Project 2001, Legal Framework for the Commercial Use of Outer Space, Working Group on National Space Legislation, Draft Report at 10 [unpublished].
376 Outer Space Treaty article XI.
377 *Ibid.*
378 Brownlie, *supra note* 104 at 444.

space by States is limited by analogous rights of other States. Manfred Lachs has put forward that "the freedom of action of States in outer space [...] is determined by the right and interest of other States [and] can therefore be exercised only to the extent to which it does not conflict with those rights and interests."[379] The doctrine of abuse of rights was recognized by the International Court of Justice in the Anglo Norwegian Fisheries case.[380]

There are several manifestations of the abuse of rights which derive from the principle of good faith[381]. These include the fictitious exercise of a right, which forfeits a right when it is exercised for the sole purpose of causing injury to another[382], and the malicious exercise of a right, which disregards the form of the law when it is used to cover the commission of what in fact is an unlawful act[383]. A clear example of the abuse of rights principle would be a complete militarization of space by a State or group of States. Even if militarization of outer space as such is not prohibited[384] a situation where one State or group of States vastly militarized outer space would be a clear abuse of right and thus invalid under International Law.

It may be concluded from the foregoing that whenever a national activity in outer space may be considered to fall within the scope of the doctrine of the abuse of rights or any of its manifestations States should refuse the authorization required by article VI of the Outer Space Treaty.

17. LIMITS FOR THE ADOPTION OF NATIONAL LEGISLATION

International Law imposes important limitations to the States in the exercise of their power to authorize activities in outer space. These limitations include the abuse of discretion and the interdependence of rights and obligations.

---

379 Lachs, *supra* note 246 at 117.

380 Anglo Norwegian Fisheries (1951) ICJ: Norway/UK.

381 B. Cheng, *General Principles of Law as Applied by International Courts and Tribunals* (Cambridge: Grotius Publications, 1987) at 121 [hereinafter "General Principles"].

382 *Ibid.* at 122.

383 *Ibid.* at 121.

384 Ramey, *supra* 364 note at 1.

18. ABUSE OF DISCRETION

It has been argued that all the ambiguities contained in article VI of the Outer Space Treaty create discretion at the national level for the implementation of domestic space law[385]. However, in international law there is a clear limit to the powers of States in the domestic exercise of their international obligations, which has been embodied in the doctrine of the abuse of discretion. According to Bin Cheng "whenever a state enjoys a certain discretionary power it must exercise it in good faith, which means that it must exercise it reasonably, honestly, in conformity with the spirit of the law and with due regard to the interests of others[386]".

For example, in the event that a national space law imposed excessively burdensome requirements for the authorization of an activity in outer space or excessive fees for the issuance of the authorization which would deter many companies to apply for an authorization that national law would be banned by the doctrine of abuse of discretion.

Therefore, all the issues which the International Space Law instruments refer in a broad manner to the states for implementation at the domestic level, as well as the ambiguities of these texts which give ample discretion to the States, are limited by the doctrine of the abuse of discretion and therefore States may not adopt national measures that would violate the spirit of the Outer Space Treaty.

19. INTERDEPENDENCE OF RIGHTS AND OBLIGATIONS

When a State agrees to abide by an obligation under an international treaty, any right which that State may have, including a sovereignty related right, that contravenes the assumed international obligation is limited and even superseded by that obligation[387]. This doctrine, known as interdependence of rights and obligations[388] imposes a clear limit to States in the implementation of domestic space law or any other national space measure. In effect, regardless of the internal constitutional, legislative or judicial prescriptions when a State undertakes an obligation at the international level the State may

[385] "Fitness of International Space Law Instruments", *supra note* 115 at 5.
[386] "General Principles", *supra* note 381 at 133.
[387] "General Principles", *supra* note 381 at 123.
[388] This doctrine was recognized by the Permanent Court of Arbitration in the North Atlantic Coast Fisheries Act. 1 H.C.R., at 169.

not adopt a national measure in contradiction with the international obligation, for that would be in itself a violation of International Law.

This doctrine, which was recognized by the Permanent Court of Arbitration as a principle of International Law, also finds support in the Vienna Convention on the Law of the Treaties which prohibits a State to "invoke the provisions of its internal law as justification for its failure to perform a treaty"[389].

Publicists concur that the relationship between international obligations of States and domestic law is well settled and the international courts [390] have consistently upheld the fact that a provision of national laws may not prevail over those of an international treaty[391]. As held by the Permanent Court "a State cannot adduce as against another State its own Constitution with a view to evading obligations incumbent upon it under international law or treaties in force."[392] It is thus submitted that as a matter of domestic law, States should take into account their international obligations when they implement an authorization procedure or grant a license.

---

[389] Vienna Convention, article 27.
[390] Brownlie, *supra note* 104 at 37.
[391] *Ibid.* at 37.
[392] Polish Nationals in Danzig (1931), P.C.I.J., Ser A/B, no. 44 at 24, quoted by I. Brownlie, Brownlie, *supra note* 104 at 37.

C. REGISTRATION

The *Corpus Juris Spatialis* has also devised a mandatory system for the registration of objects, which provides for both substantive and procedural obligations for States to implement at the national level. It is thus necessary to briefly examine the registration provisions contained in the international space law instruments to find out the extent of the obligations and rights of the States vis-à-vis the adoption of national space measures in this area.

1. NATIONAL REGISTRY

The Outer Space Treaty establishes that "a State on whose registry an object launched into outer space is carried shall retain jurisdiction and control over such object, and over any personnel thereof, while in outer space or on a celestial body"[393]. The Registration Convention elaborated upon this principle and structured a dual system of national and international registration[394]. The Registration Convention obliges States to create a national registry. In this respect, it prescribes that "when a space object is launched into earth orbit or beyond, the launching State shall register the space object by means of an entry in an appropriate registry which it shall maintain[395]".

As with the implementation of the authorization and continuing supervision principle, this does not necessarily imply the obligation to pass domestic law to create this registry[396]. Therefore, States are free to implement the registry by means of several legal mechanisms[397]. The Registration Convention also contains norms which specifically remit certain formal and procedural issues to the States. In this respect, it sets forth that "the contents of each registry and the conditions under which it is maintained shall be determined by the State of registry concerned.[398]" Therefore, there are no

---

393 Outer Space Treaty, article VIII.
394 A. A. Cocca, "Registration of Space Objects", in "Manual on Space Law", *supra note* 145 at 180.
395 Registration Convention, article II 1.
396 See *supra* note 210.
397 P. Dann, "The Future Role of Municipal Law in Regulating Space Related Activities", in T. L. Zwaan, ed., Space Law: Views of the Future (Deventer: Kluwer Law and Taxation Publishers, 1988) at129. These would include, for example, the extension of existing national registries, such as the aircraft registry, to incorporate the registration of space objects and the adaptation of court-operated registries, such as commerce registries, to include space objects, which do not imply the enactment of national space legislation.
398 Registration Convention, article II 3.

international constraints with respect to the form of the national registry. Thus, States can create a new independent registry, can incorporate the space object registry within other national registries or even require a private organization to operate the registry[399].

The Registration Convention has not determined the moment when States must create the space registry. All it says is simply that each launching state has to inform the Secretary-General of the United Nations of the establishment of the registry. This has led to the interpretation that States may establish their registries whenever they are in a position to use the registry on a frequent basis[400].

2. INTERNATIONAL REGISTRY

One of the most salient features of the Registration Convention is the fact that it created an international registry for the first time in history[401]. The Registration Convention also mandates States to communicate certain information regarding the space object recorded at the national level to the Secretary-General of the United Nations, who maintains a central registry where the information furnished by the States is recorded.[402] The Convention makes no specific reference to the time in which the information is to be given, i.e., before or after the launch. Since the Convention is silent on this

[399] In the United States, section. 415.10 of the Code of Federal Regulations establishes domestic US requirements to comply with the provisions of the Registration Convention. In this respect, it is prescribed that each licensee of a launch service is responsible for registering all objects placed in space in the course of conducting activities authorized by its license, except for objects owned by a foreign entity. Registration of objects owned by a foreign entity is the responsibility of that foreign entity. Within 30 days after launch, each licensee has to submit to the Office of Commercial Space Transportation the following information concerning any vehicle or other object it has launched into outer space: (i) the international designator of the space object(s); (ii) date and location of launch; (iii) basic programmed orbital parameters, including: (a) nodal period, (bi) inclination, (c) apogee; and (iv) General function of the space object.

[400] See rationale for the implementation of the registry in Argentina *infra* chapter III.

[401] A. A. Cocca, "Registration of Space Objects", in "Manual on Space Law", *supra note* 145 at 180.

[402] There is full and open access to the information of this central registry. The Convention is an improvement over the voluntary registration system not only because it makes registration mandatory but also in view of the fact that it provides for uniformity in relation to the data to be reported. S. Gorove, "Convention on Registration of Objects Launched into Outer Space. Analysis and Commentary" (1976) 19 *IISL* at 292.

matter, States are free to decide on the time to make the notification to the Secretary-General of the United Nations. However, in light of the object and the purpose of the Convention, the notification, if possible, should be made prior to the launching.[403]

3. PURPOSE

The main purpose for the registration of the object in the national registry is to secure jurisdiction and control over that object in outer space[404]. By recording an object in a national registry, the launching state acquires the status of State of registry, which triggers off the jurisdiction and control consequences spelled out in article VIII of the Space Treaty[405]. Thus, absent an agreement to the contrary[406], the registration allows a State to make national laws applicable to both the space object and its personnel[407].

---

[403]A. A. Cocca, "Registration of Space Objects", in "Manual on Space Law", *supra note* 145 at 180. However, Maiorsky contends that these critics often seem to be under the influence of analogies with the system of registration of aircraft, marine vessels and automobiles, which are of a different and formal nature and require the mandatory marking of registered objects. For Maiorsky, direct analogies in this case are incorrect and out of place, for while the identification of marked aircraft or automobiles may be achieved with the naked eye or with the aid of a simple optical device the same cannot be expected from the registration of space objects· In our opinion, Maiorsky's standpoint reflects the opposition of the former Soviet Union to the identification of space objects, for any mark on the space object facilitates the identification of the object, which leads to the launching state, therefore permitting the attribution of liability. B. Maiorsky, "Does the Convention on Registration of Objects Launched into Outer Space Require Revision?, (1985) 28 *IISL* at 179.

[404] Ferrer Derecho Espacial, *supra* note 323 at 282.

[405] Outer Space Treaty, article VIII.

[406] In the case of a joint endeavor, a state of registry may reach an agreement with the rest of the launching states for the application of a certain area of the law of a state other than the state of registry. The International Space Station Agreement represents the most significant instrument where the different launching states agreed on a specific mechanism for the exercise of jurisdiction and control. In effect, States opted for the registration in a separate way of each element contributed by the states. European states delegated this responsibility to the European Space Agency. Pursuant to article 5 of the Agreement each partner will register as space objects the flight elements which it provides, which are all listed in an annex to the Agreement. IGA.

[407]J. Hermida, "Space Registry" (1996) 24 *International Business Lawyer* at 383. When two or more states launch a space object States are also required to determine which one of them will register the object in its national registry. These States may further agree on the application of certain aspects of the legislation of the State which will not act as state of registry.

4. STATE OF REGISTRY

The state of registry has been defined in the Registration Convention as "a launching State on whose registry a space object is carried in accordance with article II". From this definition, it follows that there may be only one state of registry and that the state of registry has to be one of the launching states, i.e., a State which launches or procures the launching of a space object; a State from whose territory or facility a space object is launched[408]. In the event that there are several launching states these have to determine which one of them will register the object in its national registry. These states may further agree on the application of certain aspects of the legislation of the state which will not act as state of registry[409].

Difficulties have arisen with respect to the transfer of satellites in orbit, especially in the case of sale of satellites. In this respect, two sets of facts have to be differentiated: the transfer of satellites between launching states and the transfer of satellites to a non launching state. There have been some cases in practice, such as the transfer of satellites registered in the United Kingdom to China as a consequence of the hand over of Hong Kong or the sale of Canada's Anik CI and CII satellites to Argentina,[410] among some other ones. These isolated cases may not be considered to amount to a general and consistent practice of States followed from a sense of legal

---

[408] A. A. Cocca, "Registration of Space Objects", in "Manual on Space Law", *supra note* 145 at 180.

[409] Commercial Space, *supra* note 7 at 63. It must be highlighted that in the International Governmental Agreement on the Space Station states opted for the registration in a separate way of each element contributed by the states. European states delegated this responsibility to the European Space Agency. In effect, pursuant to article 5 of the Agreement each partner will register as space objects the flight elements which it provides, which are all listed in an annex to the Agreement. IGA, article 5.

[410] In order to establish a temporary satellite system to comply with ITU timelines Argentina purchased the Anik CI and Anik CII satellites to Telesat Canada by the Argentine corporation Paracom S.A. Both companies -Paracom and Telesat- formed the Paracomsat joint venture, which was in charge of the operation of the Anik satellites. Both Canadian satellites had to be moved approximately 30° to the East from their original position, to be situated at 76° West (Anik CI) and 72° West (Anik CII). Paracomsat leased transponder capacity not only to Argentine corporations - ATC Cable, Crónica TV, Canal 8 de Mar del Plata-, but also to Uruguayan television channels -4, 10, and 12 of Montevideo-, which sets the basis for a potential regional system. The satellites remained registered in Canada. J. Hermida, "Argentine Space Law and Policy" (1996) XXI-II *Ann. Air & Sp. L.* at 177 [hereinafter "Argentine Space Law and Policy"].

obligation, and thus there is no rule of customary international law governing the transfer of satellite ownership in orbit[411]. Since recourse to general principles does not either offer any solution to this problem the analysis must be done exclusively in light of the conventional sources.

In this respect, the legal literature agrees that transfers of satellites in orbit among launching states would be permitted under the Registration Convention and would not offer major difficulties, such as is evidenced by the Hong Kong precedent[412]. However, with respect to the transfer of ownership in orbit to a non launching state, the answer given in the literature is that the Registration Convention does not permit any modification in this case[413]. This assertion ignores the full array of possibilities permitted by general international law[414]. In this respect, the Registration Convention allows the possibility of launching states to conclude agreements on jurisdiction and control over the space object and over any personnel thereof[415]. Thus, launching states can decide to transfer certain jurisdictional rights to others, such as in the case of criminal law under the Intergovernmental Agreement on the International Space Station[416]. As put forward by Cocca, this reflects the principle that special agreements override general ones and the unitary criteria of article II.1 of the Convention[417].

Nothing in article II of the Registration Convention prohibits the transfer of all the jurisdictional and control rights. Therefore, it is legally possible for a State to register a space object and to enter into an agreement with another launching state to transfer part or all of the rights and obligations arising from the registration of a space object. Moreover, it is legally tenable to transfer rights to a non launching state, for the Registration Convention simply prescribes that any such agreement must be made among the launching states. In this respect, a launching state which intends to transfer rights to a non launching state will have to conclude with all the other launching states the transfer of jurisdiction and control rights and obligations to a third non

---

411 *Ibid.* at 178.

412 Lee, *supra* note 162 at 148. "1999 Remarks", *supra* note 160 at 309.

413 *Ibid.* at 309.

414 "Ferrer Derecho Espacial" *supra* note 323 at 282.

415 Registration Convention, article II.

416 Agreement among the Government of Canada, the Governments of Member States of the European Space Agency, the Government of Japan, the Government of the Russian Federation, and the Government of the United States of America concerning Cooperation on the Civil International Space Station, signed January 29, 1998, article 22.

417 A. A. Cocca, "Registration of Space Objects", in "Manual on Space Law", *supra note* 145 at 180.

launching state[418]. Under general public international law, the stipulation of rights to a third State is permitted both under customary and conventional law. In effect, in the Free Zones Case[419], the Permanent Court of International Justice held that "it cannot be lightly presumed that stipulations favorable to a third state have been adopted with the object of creating an actual right in its favor. There is however nothing to prevent the will of sovereign states from having this object and this effect. The question of the existence of a right acquired under an instrument drawn between other states is therefore one to be decided in each particular case: it must be ascertained whether the states which have stipulated in favor of a third state meant to create for that state an actual right which the latter has accepted as such."[420] Furthermore, article 34 of the Vienna Convention establishes that no rights and responsibilities may be created for third parties, except for with the consent of the third party. In such case, there must be acceptance of the third party. In case when the parties intended to create a benefit, the acceptance may be presumed. However, in cases when the States created an obligation acceptance must be in writing[421].

Therefore, it is submitted that the transfer of satellite ownership in orbit is legally possible under international law. This, however, requires an agreement among the launching states to transfer all of the jurisdiction and control rights and obligations in favor of a third non launching state. In light of customary and conventional international rules on effects of treaties to third parties acceptance in writing of the non launching state is essential.

5. RECORDED INFORMATION

It has widely been held that the States must record the information contained in article IV of the Convention[422] in their national registries[423]. However, a simple examination of the text of the Convention indicates that the State of registry has to report to the Secretary General of the United Nations said information about each space object carried on its registry and that States are

---

[418] Registration Convention, article II.2.

[419] Free Zones Case, France v. Switzerland, 1932, PCIJ at. 97.

[420] *Ibid.*

[421] Vienna Convention, article 34.

[422] This information consists of: (i) name of launching state or states; (ii) an appropriate designator of the space object or its registration number; (iii) date and territory or location of launch; (iv) basic orbital parameters, including: (a) nodal period, (b) inclination, (c) apogee, (d) perigee; and (e) general function of the space object.

[423] "The Term Appropriate State", *supra* note 272 at 79.

also free to provide additional information[424]. In effect, the text does not prescribe or even imply that the information contained in article IV must necessarily be recorded in the national registries. States must communicate this information to the Secretary General of the United Nations but this does not oblige them to register it at the national level. It is thus submitted that States are free to decide what information must be recorded in their national registry. This extends to any other substantive and procedural aspect of the national registry. However, in light of the abuse of discretion principle, States must refrain from imposing excessive burdens to their nationals. It is submitted that these burdens would include, for example, excessive and disproportionate recording fees, the filing of information which is difficult to obtain or which is difficult to obtain at a reasonable cost and information whose disclosure would imply the breach of contractual obligations. It is further held that the freedom of States to implement the national registry must be restricted by the purpose of the international registration norms, i.e., the attribution of jurisdiction and control over space objects and their personnel. To comply with this purpose, States do not necessitate excessive or unusual information. In this regard, the information contained in article IV of the Registration Convention may serve as general guidelines for the recording requirements at the national registry level.

---

[424] "Ferrer Derecho Espacial", *supra* note 323 at 282.

D. CONCLUSIONS

In general international law a State is responsible for any violation of its international obligations when it results from an action or inaction by the government of the State, its political subdivisions or any organ, agency, official or employee acting within the scope of authority. Therefore, a State is generally not responsible for the acts of individuals or other private entities. International responsibility is based on objective standards and the existence of damage is not a condition for the existence of international responsibility. Therefore, it acts as a tool to enforce standards of conduct imposed on States rather than as a means to allocate risks.

The regime of international responsibility for space activities deviates drastically from the general norms of international state responsibility with respect to attribution rules, for States have been assigned international responsibility for national activities in outer space carried on not only by governmental agencies but also by non-governmental entities. Under the Outer Space Treaty a State bears international responsibility for the activities over which it has the opportunity to exercise legal control, i.e., activities which are within the State's *jurisaction*, whether territorial, quasi territorial or personal.

International state liability has developed rather autonomously from the doctrine of international responsibility and it is based on the proposition that absence of wrongfulness does not preclude the compensation for damage caused by an act of a State. This doctrine has been incorporated to the Liability Convention of 1972. Although the Liability Convention does not in itself, at least explicitly, attribute international liability to the launching state for damage caused by non governmental entities, from the interplay between the responsibility and liability provisions of the Outer Space Treaty and the Liability Convention it follows that under the Liability Convention, States are liable for damage caused by the space objects of their national private entities. This applies even in the cases of States which have not ratified the Outer Space Treaty, since its responsibility principles are considered customary international law.

The Liability Convention attributes international liability to the launching state, whose definition has given rise to some concerns in the legal literature, especially with regard to the concept of procuring state. Therefore, the decision as to whether a State falls within the category of procuring state is a question of fact, which should be made on a case by case basis in light of the parameters contained in the definition of launching state. Even though article V neglected to include the procuring State among those which may be jointly liable the ample definition of launching state contained in article I of the Convention, together with the general principles of joint liability, suggests that the concept of procuring state is to be read in article V. A literal interpretation of article V would run contrary to the purpose of the whole Convention.

The Liability Convention permits special arrangements between States to redistribute their financial obligations, such as the liability regime adopted for the International Space Station. These agreements are valid only among these States and are not opposable to non participating states.

Private companies may not directly make a claim against a launching state under the Liability Convention, which will create burdensome obstacles to obtain remedy in an expeditious manner since they have to request the State of their nationality to espouse their claims or one of the States specifically contemplated in article VIII of the Liability Convention to make a claim on their behalf. In this respect, it is recommended that States adopt a specific domestic norm to relieve their private companies of the difficulties arisen from the international scenario. It is proposed that this norm be modeled on the Italian law of January 25, 1983. The non binding effect of the awards of the Claims Commission is another major flaw of the Liability Convention. As encouraged by the International Law Commission, it is recommended that States make a declaration to consider the Claims Commission's decisions binding as contemplated in paragraph 3 of the General Assembly's Resolution 2777 (XXVI).

The Liability Convention has structured a dual objective and subjective system of unlimited state liability, which implies an onerous burden for States as it is considered that liability arising from space activities is the most omni-comprehensive liability regime. Therefore, it is concluded that at the national level States must adopt a regime to protect themselves from the consequences of these activities. This protection should be twofold. First, States should structure safety laws or other measures to minimize the risks derived from the space endeavors of their nationals. Second, States should adopt national legislation or other domestic legal measures establishing that States may recover all or part of the compensation paid at the international level from the actual doer of the wrong, for otherwise, this international liability system also implies the assumption of risks and liability of non governmental entities by the States themselves. Therefore, States need to adopt a risk distribution system to reallocate these risks and liability according to their political objectives in the space arena.

Under the *Corpus Juris Spatialis* States are not obliged to pass domestic space legislation. They must, however, comply with the requirements contained in article VI of the Outer Space Treaty, which prescribe that the appropriate State must authorize and supervise on a continuing fashion the activities of non governmental entities in outer space. States are free to implement the form of such authorization and supervision, which may or may not include the adoption of national law.

The authorization and supervision principle should be read in conjunction with the responsibility provisions of the Outer Space Treaty, and, thus, the concept of national activities may not be construed to be at the entire discretion of each State. There are no exemptions from authorization and continuous supervision for any activity which falls under the scope of Article VI. The legal grounds for granting or rejecting the authorization to embark on national activities in Outer Space is the adherence –or lack of adherence, respectively- to the provisions of the Outer Space Treaty. The same applies to supervision. States may, however, adopt other parameters or standards for the authorization of a space activity, provided they do not curtail the rights derived from the freedom principle and other norms of international law.

There may be several "appropriate States" in each activity that takes place in outer space. The determination of the appropriate state must be done on a case by case basis in light of the specific characteristics of each activity. The recommended approach is to examine each activity and to identify all those States that may be concerned with it. Within the appropriate state, the question of who must grant the authorization is left to the discretion of that State.

Whenever a State is substantially involved in a space activity it is not obliged to expressly authorize and supervise that activity. In this case, its participation can be considered as the authorization and continuing supervision prescribed by article VI of the Outer Space Treaty. In light of the purpose of the authorization and supervision principle, authorization may only be granted or refused before the beginning of a space activity.

The freedom principle affords States the right to widely use and explore outer space. Since an ample and continuous access to outer space would fortify the strength of the private sector industry, it is recommended that those States which seek the encouragement and development of their private space industry as one of their space policy objectives extend this freedom to their non governmental national entities –to the maximum possible extent- by adopting a licensing system that will not impose unnecessary and excessive restrictions to private firms to access outer space, provided, of course, that they comply with international law and safety standards.

The freedom principle is not absolute and is subject to several limits expressly contained in the Outer Space Treaty and other constraints arising from general international law. The former includes the following principles and doctrines (i) benefit of mankind (ii) non appropriation, (iii) peaceful purpose, (iv) avoidance of harmful contamination, and (v) non interference. The international cooperation doctrine does not impose any obligation at the national level, as it is considered that it merely constitutes a beneficial advantage for space activities. In practice, the common benefit doctrine may not impose any serious limitation to the vast majority of space activities. The restrictions to the freedom of exploration and use principle derived from general international law include the abuse of rights principle, which prescribes that States may not exercise a right in order to cause damage to another State or in a way which may impair the rights of other States. There are several manifestations of the abuse of rights, which encompass the fictitious and the malicious exercise of a right.

International law also imposes limitations on States, which reduce their ability to adopt national legislation. These restrictions include the abuse of discretion and the interdependence of rights and obligations. The former obliges States to exercise a discretionary power in good faith, reasonably, in conformity with the spirit of a treaty, and with due regard to the interests of others. The interdependence of rights and obligations precludes a State from contravening an assumed international obligation on account of national law.

The *Corpus Juris Spatialis* has devised a mandatory system for the registration of space objects. States are free to implement the registry by means of several legal mechanisms, which do not necessarily include the enactment of space legislation. States may establish their registries whenever they are in a position to use it on a frequent basis. There are no international constraints with respect to the form and content of the national registry. States are also free to decide what information must be recorded in their national registry. If possible, States should make the notification of the required information to the Secretary General of the United Nations before the beginning of the space activity. The transfer of satellite ownership in orbit is legally possible under international law both to a launching state and to a non launching state by means of an agreement among the launching states for the transfer of the jurisdiction and all control rights and obligations. In light of customary and conventional international rules on effects of treaties to third parties, in the case of transfers to a non launching state, acceptance in writing of the non launching state is essential.

CHAPTER II

THE NATIONAL LEGAL FRAMEWORK

Introduction

This chapter is concerned with the examination of the domestic legislation of States that are engaged in the exploration and exploitation of outer space. This examination will be effectuated under a micro-comparison method aimed at identifying and highlighting the differences and similarities of the selected categories with the view to detecting common standards in the compared domestic jurisdictions[425]. The underlying premise behind this comparison is that the common denominators of all the States which enacted national legislation specifically focused on space should constitute the basis for the adoption of future national frameworks of space activities in other countries.

We will analyze the national legislation of those countries that enacted either a series of specific norms dealing with the regulation of various aspects of space activities or a single legislative instrument, which governs all space activities. Thus, those States which enacted isolated norms which deal with a single aspect or those which merely announced political goals for their space activities without actually regulating them are excluded from our analysis[426]. Also excluded are those countries which regulate space activities through their general laws but which do not have any specific norms dealing with space activities[427], as well as, those which do not regulate the activities of private and other non governmental entities.

Consequently, we will examine the national norms of the United States, the Russian Federation, Australia, the United Kingdom, Ukraine, Sweden, South Africa and Japan. Together with Argentina, whose domestic

---

[425] The comparison will be based, principally, on the examination of statutory instruments, both primary and subordinate legislation, and other government developments, such as the adoption of policy documents and space programs. Judicial decisions in the space field are scarce, and in most jurisdictions they are non existent. Therefore, practically no reference will be made to court decisions. K. Schadbach, "The Benefits of Comparative Law: A Continental European View", 16 B.U. *Int'l L.J.* at 331.

[426] These countries would include Norway, Brazil and Italy, among others.

[427] Although France could be included within these countries –even if it has some domestic norms, due to its importance we will analyze French legislation in this book.

norms we will examine in detail in the next chapter, these are the countries which COPUOS identified in its 2001 report concerning the examination of key elements of existing national space legislation.[428] However, due to the importance of their long-standing active participation in the space arena we will also analyze the norms which regulate the development of space activities in France and Canada, even if these countries have not yet adopted a comprehensive domestic space regulatory framework. Canada has had a very active participation in the space field and has relatively recent announced a political and regulatory framework[429]. In France, current regulations are a mosaic of scattered contractual, administrative and regional norms and arrangements which have been adopted for each space program. The analysis of the national norms of France, which is also in the process of adopting specific legislation focused on space[430], and Canada will help to contrast and corroborate our hypotheses regarding those States which have enacted national space legislation.

The selection of the problems to be compared will be the categories dealt with in the chapter devoted to international space law. These categories are: (i) the nature of the domestic legislation, (ii) the domestic measures adopted to deal with international responsibility and liability, (iii) the implementation of the authorization and continuing supervision principle, and (iv) the compliance with registration obligations at the national level. We will also begin the analysis of each of the selected states with the examination of the relationship between international and domestic law. With respect to the authorization principle, we will focus our analysis on space activities in general. However, we will also briefly consider the specific norms existing for the approval of satellite telecommunications in some of the jurisdictions where there is a separate sophisticated and complex licensing system for these activities, most notably the United States and the United Kingdom. With regard to the national measures dealing with international responsibility and liability, we use a tripartite categorization of risks following the model of the US Department of Transportation, which was adopted in the legal literature by some authors, such as Bender[431]. Under this tripartite classification, liability risks are divided as: (i) first-party risks or risks among participants[432],

---

[428] A/AC.105/C.2/L.224.

[429] "Commercial Space", *supra* note 7 at 33.

[430] Gerhard & Schrogl, *supra* note 52 at 20.

[431] R. Bender, *Space Transport Liability: National and International Aspects* (The Hague: Martinus Nijhoff, 1995) at 208.

[432] First party risks imply the possibility of damages to the participants' space objects, i.e., the space vehicle in case of the launch company or the payload in case of the customer, and to the participants' personnel resulting from the launch activity.

(ii) second-party risks[433] and (iii) third party risks[434]. In most jurisdictions second party and third party liability risks lead to the same results. However, in more complex and sophisticated risk sharing regimes, mainly the United States, second and third party risks are allocated differently[435]. Therefore, for the sake of consistency we will analyze all jurisdictions under this tripartite classification. In those cases where the treatment of second and third party risks are identical we will briefly mention this fact without repeating our analysis. Throughout our analysis we will focus on the opportunities afforded to the space industry and to the general public in the rulemaking process and in the implementation of the space policy from a law reform and participatory theory perspective.

Our pivotal hypothesis is that even if States have established their domestic norms differently as a consequence of their own legal and political individual characteristics there are common denominators in all these domestic jurisdictions. Specifically, we hold that all countries, particularly those actively involved in the pursuit of space activities, have implemented an authorization system, one of which pillars is the State's assurance that the activities will not entail significant safety perils. Additionally, with regard to domestic space launch legislation we postulate that any legal framework aimed at governing launch services must necessarily address the issue of the allocation of risks and assignment of liability and reallocate these risks according to the country's space policy objectives.

After our examination of the national space legislation of each of the selected jurisdictions, we will test our hypotheses concerning certain trends and developments of domestic space legislation. Our analysis will depart from the theses and postulates of recent studies on comparative national space law, which differ, or at least, differ in certain aspects, from our own hypotheses. In this respect, we will first attempt to corroborate our hypothesis on the lack of necessity to adopt new rules aiming at harmonizing authorization and supervision of space activities, and on the necessity to urge States to rectify any deviation from the principles of the international legal scenario, as well as to respect the differences between laws of different countries arising from differences in substantive concerns and political values, provided they are in conformity with the international obligations. Second, we will intend to prove that, contrary to some of the findings of the alluded studies, the distribution of

---

[433] Second party risks constitute risks to certain related entities which, although they do not participate directly in the space activity, are all the same exposed to some risks
[434] Third party risks refer to the possibility of damages caused to persons and property thoroughly unrelated to the operation.
[435] Bender, *supra* note 431 at 208.

risks and liability, in particular in space transportation, does not need harmonization since its content is the result of the space policy objectives of each country, which differs substantially according to its needs and the stage of development of its space industry. However, we will postulate that the structure and main elements of the risk allocation regime tend to converge, as this term is understood in Comparative Law, in all jurisdictions, especially in those actively involved in space activities, due to the consequences derived from international obligations assumed by States.

Finally, we will examine the viewpoint which advocates the adoption of new international rules or new interpretations of existing ones and we will contrast it with our hypothesis concerning the lack of need for these new rules.

A. UNITED STATES OF AMERICA

1. NATIONAL APPLICATION OF INTERNATIONAL LAW

As a result of the separation from Great Britain, the powers of external sovereignty passed from the Crown to the United States of America and not to the states[436]. According to article II section 2 of the US Constitution the treaty making process in the United States is vested in the Federal Executive branch, which acts with the advice and consent of the Senate[437]. The President negotiates and signs an international agreement and the Senate must give its consent for the ratification, for which purpose a two-third majority is needed. Pursuant to Article VI of the Constitution, international agreements[438] which the United States is a party to are the supreme Law of the Land, together with the Constitution itself and the laws of the United States, and therefore prevail over the law of the several states[439].

The United States follows a mixed or partly dualist system with respect to the value of international treaties in US domestic law[440]. For the purpose of their domestic application treaties are essentially grouped in two

---

[436] United States v. Curtiss-Wright Export Corp. et al. No. 98 Supreme Court of the United States. 299 U.S. 304.

[437] US Constitution, art. II, sec. 2.

[438] Customary international law is also considered the law of the Land The Paquete Habana, 175 U.S. 677, 700, 20 S.Ct. 290, 299, 44 L.Ed. 320 (1900).

[439] U.S. CONST. art. VI, 2.

[440] J. H. Jackson, "United States" " in F. G. Jacobs & S. Roberts, (eds.), *The Effects of Treaties in Domestic Law* (London: Sweet & Maxwell, 1987) at 141.

distinct categories: self-executing and non self-executing agreements. A treaty is considered non self-executing if: (a) the agreement manifests an intention that it will not become effective as domestic law without the enactment of implementing legislation, (b) the Senate in giving consent to a treaty, or Congress by resolution, requires implementing legislation, or (c) implementing legislation is constitutionally required[441]. In such cases, the international treaty necessitates domestication to have legal effect on the national plane[442]. Treaties considered to be self-executing are directly applicable and may be invoked by individuals and entities before US courts[443]. For Peter Nesgos, the provisions of the Outer Space Treaty declaring rights, such as the freedom of exploration and use principle, or establishing obligations, such as the prohibition to appropriate outer space, do not require any further legislative action and thus are self-executing. Other provisions which directly call for legislative action, such as the authorization and supervision mandate, may not be considered self-executing[444]. A similar conclusion was reached by the Senate during the hearings for the ratification of the Outer Space Treaty[445]. The possibility that a treaty may be partially self-executing has long been recognized by the US courts[446].

As arises from Chief Justice Marshall's pronouncement in the Murray v. Schooner Charming Betsy decision, [447] an Act of Congress ought never to be construed to violate the law of nations if any other possible construction remains. Thus, in principle where reasonably possible, a United States statute is to be construed so as not to conflict with an international treaty[448]. However, it is a well-settled rule of US law that if a domestic act of Congress cannot be reconciled with international law a later statute supersedes an

---

441 Restatement of the Law, Third, Foreign Relations Law of the United States, § 111.

442 Generally, the intention of the United States determines whether an agreement is to be self-executing in the United States or should await implementation by legislation or appropriate executive or administrative action. After it is concluded, often the President determines its self-executing character or decides that transformation is necessary. Additionally, Congress and even the courts may decide whether an agreement is or is not self-executing. Restatement of the Law, Third, Foreign Relations Law of the United States, § 111 Comment h.

443 J. H. Jackson, "United States" in F. G. Jacobs & S. Roberts, eds., *The Effects of Treaties in Domestic Law* (London: Sweet & Maxwell, 1987) at 141.

444 "National Law and Commercial Activities in Outer Space", *supra* note 76 at 23.

445 "US Senate Hearings 1967", *supra* note 316 at 35.

446 "National Law and Commercial Activities in Outer Space", *supra* note 76 at 18.

447 Murray v. Schooner Charming Betsy, 6 U.S. (2 Cranch) 64, 118, 2 L.Ed. 208 (1804)

448 Restatement of the Law, Third, Foreign Relations Law of the United States, § 114.

earlier international treaty[449]. Additionally, no international treaty may be given domestic effect in the United States if it is inconsistent with the US Constitution[450]. However, non-compliance with an international obligation on account of domestic law does not relieve the US of its responsibility on the international plane[451].

The US courts' attitude toward the interpretation of international treaties has often been the object of criticism by authors and commentators[452]. As has been pointed out, "the time has come to reexamine the approach taken by the Supreme Court as to the interpretation of international agreements[453].

2. NATURE OF US NATIONAL SPACE LAW

The United States was the first country to adopt national legislation to regulate outer space activities. It has done so since the so called Commercial Innocence Age. The first law was the National Aeronautics and Space Act of 1958, which created NASA and established its purpose, objectives and responsibilities[454]. During the first years of the Commercial Innocence Age, the approach followed by the US was to adapt the existing laws and regulations to space activities, mainly in the satellite telecommunications field[455].

Then the United States followed a reactive approach with respect to its domestic legislation by adopting measures as its needs arose[456]. Therefore,

---

[449] State of Missouri v. Holland, United States Game Warden No. 609, Supreme Court of the United States, 252 U.S. 416; 40 S. Ct. 382; 64 L. Ed. 641; 11 A.L.R. 984; 18 Ohio L. Rep. 61.

[450] Reid, Superintendent, District of Columbia Jail, v. Covert No. 701, October Term, 1955, Supreme Court of the United States, 354 U.S. 1; 77 S. Ct. 1222; 1 L. Ed. 2d 1148. "No agreement with a foreign nation can confer power on the Congress, or on any other branch of Government, which is free from the restraints of the Constitution."

[451] Vienna Convention on the Law of the Treaties, article 27.

[452] D. J. Bederman, Revivalist Canons and Treaty Interpretation, (1994) 41 *UCLA L. Rev.* at 953.

[453] M. A. Rogoff, "Interpretation of International Agreements by Domestic Courts and the Politics of International Treaty Relations: Reflections on Some Recent Decisions of the United States Supreme Court" (1996) 11 *Am. U.J. Int'l L. & Pol'y* at 567.

[454] National Aeronautics and Space Act of 1958 [hereinafter "NASA Act"].

[455] In 1970 the FCC declared the Communications Act of 1934 applicable to space telecommunications. Afterwards, many specific satellite telecommunications regulations were adopted.

[456] N. C. Goldman, *American Space Law International and Domestic* (Ames: Iowa

as will become apparent from our discussion, the US does not have a single or primary legislative instrument that governs its national space activities.[457] Thus, US national space law consists of a series of laws and regulations which govern specific aspects of different space activities, as well as several non specific norms which have a direct impact on the space industry, such as export and tax laws. For example, in the commercial launch industry, in 1983 the United States government initiated a regulatory process by enacting a series of instruments aimed at fostering the private sector industry[458]. This regulatory movement began with the National Security Decision Directive No. 94 issued in May, 1983, and was further developed with the adoption of (i) the Executive Order 12465 of February 24, 1984, whereby the Department of Transportation was designated as the lead agency in commercial space launch activities[459], (ii) the National Policy on the Commercial Use of Space of July 20, 1984, which provided certain tax benefits to the space sector, and (iii) the Commercial Space Launch Act of 1984, which simplified the launch licensing procedure[460].

In the satellite telecommunications sector, the Federal Communications Commission, which has been entrusted with the allocation of radio frequencies pursuant to the Federal Communications Act of 1934[461], has been promoting the adoption of a sophisticated licensing procedure and norms for access to the satellite telecommunications market, as well as regulations for the operation of satellites in all orbits[462]. In the satellite remote sensing field, the regulatory regime suffered from the constant changes and improvisation in the satellite remote sensing policy and the privatization projects[463]. Only with the adoption of the Land Remote Sensing Policy Act of

---

State University Press, 1988) at 23.

[457] A/AC.105/C.2/L.224 at 13.

[458] Hermida, J., "Risk Management in Commercial Launches", (1997) *Sp. Pol.* at 13 [hereinafter "Risk Management in Commercial Launches"].

[459] Exec. Order No. 12,465, 49 Fed. Reg. 7211 (1984).

[460] Although all these legal measures intended to promote the industry, they did not succeed in fostering commercial launch providers to offer their services to the users. As mentioned above, the reasons of this failure derive, basically, from (i) the fact that these legal measures did not include a risk management system which could allow launch companies to transfer risks and to limit the liability, which under national and especially under International Space Law constitutes a rather onerous burden, and (ii) the conditions under which NASA offered the Space Shuttle services, which impeded the private space launch operators to compete with the Administration. "Risk Management in Commercial Launches", supra note 458 at 13.

[461] 47 USC 151.

[462] *Ibid.*

[463] "Commercial Space Law", supra *note* at 216.

1992 (and its several amendments) did the US manage to implement a clear set of rules for the remote sensing industry[464].

Due to the importance of the US space industry, many non specific norms and procedures have a direct impact on the regulation of commercial space activities. For example, the US export laws limit and condition the export of satellites and their component parts to a select group of nations[465]. Also, specific tax provisions have been enacted to provide the space industry with tax tools which take into account the specific features of space technology and its needs[466].

Another significant characteristic of US space domestic law is the importance of a great number of regulations emanated from administrative agencies, such as the Department of Transportation, NASA, National Oceanic and Atmospheric Administration, and the Federal Communications Commission. These regulations aim at implementing the legislative directives contained in the major acts and regulating in detail all aspects related to space activities.

An essential feature of the regulatory process is the Public Notice system, whereby a federal agency publishes in advance a proposed regulation for comments from the public, then it debates and analyzes those comments and finally it enacts the proposed regulation in light of the results of those comments and analysis.[467] This system derives from the due process and sound governmental policy principles, which require giving the public an opportunity to participate in rulemaking proceedings that might affect private business interests or the personal liberties of private citizens[468]. This process allows ample participation of all those interested in the formulation of the regulations. In particular, the rulemaking process of the Federal Aviation Administration, which has responsibility to enact regulations under the Commercial Space Launch Act[469], specifically takes into account the industry

---

464 *Ibid.* at 217.

465 R.S. Jakhu & J. Wilson, "The New US Export Control Regime and its Impact on the Communications Satellite Industry", (2000) XXV Ann. Air & Sp. L. at 157.

466 "Case Study", *supra* note 65 at 125.

467 Federal regulations and other agency rules ordinarily must be promulgated in accordance with the public notice-and-comment procedures specified by the Administrative Procedure Act ("APA"). R. A. Anthony, "Interpretive" Rules, "Legislative" Rules and "Spurious" Rules: Lifting the Smog, 8 Admin. *L.J. Am. U.* at 2.

468 Tenn. Juris., Administrative Law, § 3

469 The Secretary of the Department of Transportation delegated these responsibilities

needs and incorporates the experience of the industry gained through the operation and administration of space activities and the results of research and development activities to maintain the regulations current and useful with respect to the industry operational needs[470].

3. RESPONSIBILITY AND LIABILITY

The 1988 Amendments to the Commercial Space Launch Act introduced a system for the allocation of risks and assignment of liability between private participants and the government, and also between private participants themselves. This system permitted the US private industry to offer services at a competitive cost[471].

3.1. First party risks

First party risks are allocated through a system of mandatory reciprocal waivers of liability. These waivers of liability, which have been modeled after the ones used by NASA[472], represent legislatively mandated contractual indemnification obligations of each private participant, and its contractors and subcontractors, vis-à-vis the other private participants, as well as its contractors and subcontractors.

In order to avoid claims among the different participants in a space launch, and as a condition to the issuance of the license, commercial launch providers are required to "enter into reciprocal waivers of claims with their contractors, subcontractors and customers, as well as the contractors and subcontractors of such customers involved in launch services, by virtue of which each party to each such waiver agrees to be responsible for any property damage or loss it sustains or for any personal injury, death of, or property damage or loss suffered by its own employees resulting from

---

to the FAA, which in turn assigned them to the Associate Administrator for Commercial Space Transportation 14 CFR Chapter III.

[470] W. A. Gaubatz, "International Certification for Commercial Reusable Space Transportation", (1999) 42 *IISL* at 257.

[471] "Space Risk Management", *supra* note 67 at 48. The analysis of the distribution of risks and assignment of liability in the United States, Australia and France profits substantially from our analysis carried out in our LL.M thesis. J. Hermida, *Legal Aspects of Space Risk Management. The allocation of risks and assignment of liability in commercial launch services*, (LL.M., Thesis, McGill University, 2000) [unpublished].

[472] *Ibid.*

activities carried out under the use of the license."[473] As can be observed from the quoted provision, the waiver of liability requires that each party, as well as its contractors and subcontractors, (i) be responsible for damages it sustains, and (ii) refrain from making claims against the other party, and its contractors and subcontractors.[474]

According to the text of the 1988 Amendments, the purpose of this provision is (i) to limit the total universe of claims that might arise from a launch, and (ii) to eliminate the necessity for all the parties to obtain property and casualty insurance to protect against such claims.[475] With respect to the first of the objectives sought by the 1988 legislator, as the reciprocal waivers promulgate the assumption of risks by each participant they act as a deterrent of claims. Indeed, by virtue of this legal prescription each participant is precluded from its right to sue the entity causing the damage[476]. The scope of this provision does not encompass all events which may originate damage arising from a space launch. Moreover, even within the covered events not all claims are precluded. However, the waivers have proved to act as an effective hindrance of lawsuits.

As regards the second of the objectives sought by the legislator, the waivers of liability foster the obtainment of insurance –or another form of risk management- by the users to protect against their own first party risks, for they may not afford to lose their payload without recouping at least part of their investment. However, since neither the launch carrier nor the customer is liable for damage it may cause to each other, the obtainment of insurance to protect against foreign first party risks becomes thoroughly unnecessary.[477]

Therefore, it is submitted that the actual objective of the reciprocal waivers of liability consists of providing the launch industry with a system that permits it to convey risks to the customers. In effect, notwithstanding the

---

[473]49 USC §2515

[474] P. Meredith, "Risk Allocation Provisions in Commercial Launch Contracts", (1991) 34 *IISL* at 267 [hereinafter "Risk Allocation"].

[475]Commercial Space Launch Act Amendment of 1988, Report of the Senate Committee on Commerce, Science and Transportation on H.R. 4399, SR 100-593, Oct. 7, 1988, US Government Printing Office, Washington, 1988, at 14 [hereinafter "Senate Report"].

[476]"Risk Allocation", *supra* note 474 at 267.

[477] In this respect, the waivers executed by private entities are intended to relate only to risks normally covered by insurance or self-insurance. For an analysis of the concept of foreign first party risks, see "Space Risk Management", *supra* note 67 at 68.

alleged objectives of the Act, the main purpose of these waivers is to give the launch services provider the possibility of not having to compensate the damage it causes to its customers and its contractors and subcontractors. In practice, this operates as an exclusion of liability. The rationale of this legal provision is that the launch provider transfers to the user the damage the latter may suffer as a consequence of a failed launch or any other event, whether attributable or not, to the launcher. Even if at the same time, the user transmits the launch operator the risks which the operator might sustain as a result of the user's conduct, in practice this hypothesis is extremely unusual. Consequently, the waivers act only as a mere scheme enabling the transfer of risks from the launch operator to the user of the space transport services. This exclusion of liability constituted the only viable alternative for the emergence of the US private launch industry at the time of the enactment of the 1988 Amendments to the Commercial Space Launch Act[478].

With respect to the damages covered by the waivers of liability, apart from damages to the employees, these waivers encompass the obligation to assume only property damage and losses. It may thus be concluded that indirect damages and consequential damages are outside the scope of the waivers. As regards the damages to the employees, the Act circumscribes these damages to the cases of personal injury, death, and property damage or loss[479]. To Peter Nesgos, the concept of personal injury must be construed exclusively as bodily injury.[480] However, when the legislator intended to restrict this concept to cover only bodily injuries she did so specifically as, for example, in the case of governmental indemnity. Therefore, it is submitted that the reciprocal waivers of liability include not only bodily injuries but also other non bodily damages, such as mental trauma, nervous shock, mental anguish, and psychic trauma.[481] Additionally, the FAA in its Final Rule adopted on October 26, 1998 clarified that the term "bodily injury means physical injury, sickness, disease, disability, shock, mental anguish, or mental injury sustained by any person, including death", thus putting an end to any

---

[478] *Ibid.* at 85.

[479] 49 USC §2515.

[480] P. Nesgos, "Recent Developments in Risk Allocation of Concern to the US Commercial Space Launch Industry and the Insurance Community", (1989) Assicurazioni Generali, Fifth International Conference on Space Insurance at 16 [hereinafter "Recent Developments"].

[481] For an analysis of these concepts in the aviation industry see, Georgopoulos v. American Airlines South Wales, 1993, T.T. Burnett and Winifred Burnett v. Trans World Airlines, Inc United States District Court of New Mexico, 1973, Rosman et al v. Trans World Airlines, Inc. New York Court of Appeals, June 13, 1974, Husserl v. Swiss Air, United States District Court, Southern District of New York, 1975, Eastern Airlines, Inc. v. Floyd et al. Supreme Court of the United States, April 17, 1991.

other possible construction of this term.[482] The existing case law appears to indicate that damages which are caused through gross negligence or willful misconduct are not covered.[483]

3.2. Second party risks

Risks to the government property are distributed in a two-layered basis, where the private launch operator assumes the risk of losses through a system of insurance or self-insurance up to the amount of the maximum probable loss, which may not exceed US$ 100,000,000 or the maximum liability insurance available on the world market at reasonable cost if the amount is less than US$ 100,000,000. The government absorbs the risks from that limit upwards through the so called waivers of liability.

Thus, the Commercial Space Launch Act obliges launch operators to obtain liability insurance or to demonstrate financial responsibility in an amount sufficient to compensate the maximum probable loss from claims against any person filed by the United States for loss of or damage to property of the United States resulting from activities carried out under the license in connection with any particular launch[484].

---

[482] Department of Transportation Federal Aviation Administration 14 CFR Part 440 Financial Responsibility Requirements for Licensed Launch Activities. Final rule, October 26, 1998 [hereinafter "Final Rule"].

[483]S. Gorove, *Cases on Space Law*, (Mississippi: Journal of Space Law, Inc., 1996) at 1.

[484] The limitations included in the 1988 Amendments are based on the best assessment received by the Committee of a maximum probable loss to US property. The biggest vehicle of the commercial fleet existing at the moment of the research carried out for the Committee, was Titan III, manufactured by then Martin Marietta. The worst accident of this vehicle had caused damages to two launch bases, which were out of use for 9 months. Total losses for this accident amounted to US$ 60,000,000. Therefore, it was felt that requiring a limitation of US$ 100,000,000 was reasonable. During the first years of commercial space launch activities following the enactment of the 1988 Amendments the Commercial Space Transportation Office determined that the maximum probable damage averaged US$75,000,000 to US$ 80,000,000 for expendable launch vehicles such as Delta, Titan and Atlas Centaur launched from Cape Canaveral, Florida and US$ 100,000,000 for vehicles which make suborbital launches from White Sands, New Mexico.
D. E. Cassidy, "Allocation of Liabilities Between Government and Private Sector and Implications on Insurance for Space Commercialization", (1990) 33 *IISL* at 28 [hereinafter "Allocation of Liabilities"].

The maximum probable loss does not cover all the damages which may arise in a launch, but only those which may take place in the majority of accidents related to space launches[485]. It was the opinion of the Senate Committee that it would be contrary to the public interest of the United States to permit a launch from governmental bases without requiring an adequate insurance regime that protects the assets of the United States. Even if aware that due to the fluctuations of the market it may not be possible to obtain insurance against the US property in an amount sufficient to achieve the maximum probable risk standard, the Senate Committee understood that such circumstances were infrequent and that the damages that may occur which exceed the insurance will be almost insignificant. In such cases, the Senate Committee understood that the Department of Transportation should require the licensee proof of financial responsibility in an amount that covers the difference between the available insurance coverage and the maximum probable loss. If even so, this combination between insurance and financial responsibility are still insufficient, the Senate Committee was of the opinion that the Secretary should have discretionary authority to nonetheless issue the license. It can be gathered from the above, that there are certain risks to US government property which are assumed by the US government itself, rather than the space launch carrier[486].

The United States, its agencies, contractors, and subcontractors involved in launch services are obliged to enter into reciprocal waivers of claims with the commercial launch provider, its contractors, subcontractors and customers, as well as the contractors and subcontractors of such customers[487]. These waivers only apply to the extent that the claims exceed

---

[485] The beneficiaries of the insurance or the demonstration of the financial responsibility are the United States, its agencies, contractors and subcontractors, personnel and the customer of the launch licensees, and its personnel, without any cost to the United States. Currently, the determination of the maximum probable loss is done pursuant to Appendix I of the Final Rule, which contains information requirements that a launch carrier must submit to the FAA. This information includes (i) general information regarding the description of the mission, the launch vehicle, the payload and the flight termination system, (ii) pre-flight processing operations, (iii) flight operations and (iv) post-flight processing operations. "Final Rule", *supra* note 482.

[486] Like in the waivers of liability among the private sector entities, the issue of the possibility of filing claims to enforce contractual remedies has arisen. This may take place, for example, in the event that the Air Force or NASA undertakes to offer certain alternatives in cases of breach of the arrangement to provide space launch facilities. In this regard, the same conclusions apply, i.e., nothing in the statute precludes the exercise of such rights and obligations.

[487] By virtue of these waivers of liability each party to each such waiver agrees to be

the amount of property insurance or demonstrated financial responsibility required as a condition of the license. In the event that the launch provider has opted to obtain insurance, the government will directly receive the proceeds of such insurance policies that are paid following an accident that causes damage to the government property[488].

---

responsible for any property damage or loss it sustains or for any personal injury to, death of, or property damage or loss sustained by its own employees resulting from activities carried out under the license.

[488] With respect to public sector participants, the Government acting through the Secretary of Transportation enters into the waiver of liability agreement on behalf of its contractors and subcontractors, who do not therefore sign this agreement. As explained in the Final Rule of October 26, 1998, the FAA views Government contractors and subcontractors as third-party beneficiaries of the reciprocal waiver agreement and the Government is responsible for protecting their interests. The implication of the waiver of liability for claims entered into by the US government on behalf of its contractors and subcontractors is that the Government assumes the risk of their property damage exceeding the amounts of the insured policy. According to the FAA, the Government manages this risk in a twofold way: "First, the licensee is required to obtain property insurance covering damage or loss to property of Government contractors and subcontractors involved in licensed launch activities, in addition to Government-owned property. Second, Government contractors and subcontractors must also maintain insurance for their property, the cost of which is charged to the Government as an allowable cost. In the event Government contractor property is damaged, the Government would look first to the licensee's property policy for coverage in order to relieve financial risks to the Government. The contractor's insurance would cover the second tier of risk up to policy limits. In both instances, the risk of loss above statutorily-required insurance is borne by the Government." In situations where the scope of property insurance coverage available is limited by policy exclusions, such as war risk, workers' compensations, radio wave interference or environmental hazards, the amended Act permits the government to extend the waiver of claims to those excluded areas. In effect, the Secretary of Transportation is entitled to execute, on behalf of the United States and any of its agencies, waivers of liability in the event that there is no insurance coverage available on account of exclusions deemed usual in the market. This practice implies that in these cases the government assumes all the damages, i.e., from the first dollar without any limit. For example, this would be the case of damages to government launch facilities caused by a space launch provider attacked by a missile or another weapon during a war. In such case, the damages would be entirely assumed by the government since they were not covered by the insurance. It must be pointed out that the assumption of these risks is not mandatory for the United States. The Secretary of Transportation has been empowered with amply discretionary authority to decide, after a consultation process, whether or not to assume liability on behalf of the United States in cases of insurance exclusions. While we understand that this provision is both necessary and beneficial for the launch industry, we are of the opinion that the language used in the Act is detrimental for those launch providers that opt to self-

The general requisite of the reciprocal waivers is that each private sector participant waives its right to make claims to public sector participants, and these in turn are to waive their rights to sue the private sector participants. The chain of private participants is limited by their involvement in the space launch.[489] The success of the system of the waivers of liability is achieved when all participants, even those indirectly related, agree to be bound by the waivers. Otherwise, the objective sought by the Act may be seriously affected[490].

3.3. Third party risks

Third party risks are allocated between the private launch provider and the government on a horizontal basis, consisting of three layers. In each of these layers, either the government or the launcher assumes the risks. In the first layer, the launch carrier assumes all risks up to the maximum probable loss, which may not exceed US$ 500,000,000 or the maximum liability insurance available on the world market at reasonable cost if the amount is less than US$ 100,000,000. Instead of procuring insurance the carrier may resort to demonstrate financial responsibility, i.e., self-insurance or any other risk management alternative. The possibility to resort to self-insurance is a response to the space insurance market conditions prevailing in the 1980's.

---

insure the risks instead of resorting to obtain insurance in the market.

489 Thus, for example, when the user is a satellite telecommunications company this company, the manufacturer of the satellite, the manufacturers of the satellite components, whether these make the components at the request of the manufacturer (contractors) or at the request of an entity which the manufacturer hired for the manufacture of a component (subcontractors), are among the entities which the government is obligated to enter into reciprocal waivers with. Also included are the manufacturer of the space vehicle, when it is not the carrier itself, the manufacturers of the component parts of the vehicle, both under a direct contract with the manufacturer and with the contractors, and the satellite ground transport company which carries the satellite from the manufacturer's premises to the launch base, whether under a contract with the launch carrier, manufacturer or user. "Allocation of Liabilities", *supra* note 483 at 27.

490 Another salient aspect of these provisions is that the public sector entities are not required to enter into reciprocal waivers of liability among themselves. Thus, for example, the Air Force may sue a contractor which caused damages to a federal range. Additionally, according to the FAA rules of October 26, 1998, the US government employees are not required to waive their claims under the reciprocal waiver of claims agreement. Therefore, any injured employee is free to elect to seek compensation from a negligent launch participant from whom he or she does not work or to exercise a labor claim against his or her employer "Private Launch", *supra* note 30 at 158.

The limitation of the insurance to be obtained to the maximum probable loss was also a response to the insurance market conditions existing at the time of the enactment of the 1988 Amendments[491]. As pointed out by Valérie Kayser, the lack of definition of maximum probable damage in the Act and the ambiguity of the concept vest the Secretary of Transportation with ample discretionary powers to determine the maximum probable damage. This situation presented a possible source of problems for the industry, for the Act does not specifically foresee the possibility of challenging this determination before the Courts. [492]

The second level of the third-party risk system is made up of government indemnity, where the government absorbs all successful claims exceeding the first layer up to US$ 1,500,000,000 [493]. It constitutes the central

---

[491] In this respect, damages arising from a space launch may be classified in possible and probable. The former refer to the highest possible damages which an extraordinary and quite unlikely accident may cause. These might include, for example, damages to the entire population of a city. The actual maximum possible damages exceed the insurance -and self-insurance- capacity, at least at reasonable costs, existing in the world space insurance market. On the other hand, the probable damage is that ordinary damage which may occur in most of the accidents related to space launches. The maximum probable damage refers to ordinary accidents which may originate maximum losses. Under normal insurance market conditions, probable damages may be insured at reasonable costs.

[492]"Private Launch", *supra* note 30 at 154. The beneficiaries of the insurance are: (i) the United States Government, its executive agencies and personnel, contractors, and subcontractors of the Government and (ii) private sector participants, such as contractors, subcontractors, and customers of the licensee and the contractors and subcontractors of the customer. The inclusion of the US government may not generate any cost to the United States492. Previously, risks to government employees were considered by the industry as second-party risks and thus covered by the government indemnification. This meant that the government assumed responsibility for losses sustained by their own employees through the reciprocal waiver of liability. As we have seen, these waivers operate as from the amount of the maximum probable loss, which may not exceed US$ 100,000,000 or the highest amount of liability insurance available. Now, under the new regulation, since they are considered third parties, the government will only assume the risks for their employees as from the maximum probable loss for third party claims, which may not exceed US$ 500,000,000 or the maximum liability insurance available and only up to US$ 1,500,000,000. This clearly shows that the industry is deprived of a protection which it used to enjoy. Notice, however, that the Government would continue to be responsible for employees' claims in the event of a policy exclusion considered usual for the type of insurance in question.

[493] The relevant part of the Act establishes that "To the extent provided in advance in an appropriation law or to the extent additional legislative authority is enacted

element of the system of distribution of risk and assignment of liability between the private sector launch industry and the government. This materialized one of the concerns of the launch industry. The government thus shares with the private sector the risks which may arise from a commercial space launch.

The third layer includes all claims above the upper limit of the government indemnification, i.e., US$ 1,500,000,000 over the amount of prescribed third party liability insurance, and constitutes the exclusive responsibility of the launch provider. Risks above the US$ 1,500,000,000 level are considered possible but extremely improbable. These risks are thus referred to as maximum possible losses or catastrophic losses[494]. The assumption of risks by the launch carrier does not preclude United States' responsibility under International Law.

4. IMPLEMENTATION OF THE AUTHORIZATION AND SUPERVISION PRINCIPLE

4.1. The Commercial Space Launch Act of 1984

---

providing for paying claims in a compensation plan submitted under subsection (d) of this section, the Secretary of Transportation shall provide for the payment by the United States Government of a successful claim (including reasonable litigation or settlement expenses) of a third party against a licensee or transferee under this chapter, a contractor, subcontractor, or customer of the licensee or transferee, or a contractor or subcontractor of a customer, resulting from an activity carried out under the license issued or transferred under this chapter for death, bodily injury, or property damage or loss resulting from an activity carried out under the license. However, claims may be paid under this section only to the extent the total amount of successful claims related to one launch— A. is more than the amount of insurance or demonstration of financial responsibility required under section 70112(a)(1)(A) of this title; and B. is not more than $1,500,000,000 (plus additional amounts necessary to reflect inflation occurring after January 1, 1989) above that insurance or financial responsibility amount."

[494] It must be pointed out that the likelihood of any incident that would produce claims in excess of $1,500,000,000 is very remote. Actually, there has never been a successful third party claim in the history of the US Space program. Thus, in practice the government assumes the most burdensome role in the risk allocation system established in the 1988 Amendments. Furthermore, this system also provides the certainty to the launch operators that they will not be exposed to unlimited liability, which enables the US private-sector launch carriers to compete in better conditions with European and other foreign launchers.

The US private sector launches are mainly regulated by the Commercial Space Launch Act enacted in 1984.[495] The purpose of the Act is to "encourage the United States private launch vehicles and associated launch services by simplifying and expediting the issuance and transfer of commercial launch licenses and by facilitating and encouraging the utilization of government-developed space technology." [496] The Act is based on the premise that the development of commercial launch vehicles and associated services would enable the United States to retain its competitive position internationally. Since the adoption of this act, the U.S. private launch sector has had the possibility of offering space transport services directly to its customers[497]. For some scholars, the minimal regulation philosophy underlying the Act is bound to encourage and facilitate private enterprise in space[498].

The Act structures the authorization and supervision of space transport by means of a system of licenses needed both to carry out space launches, as well as to operate launch sites[499]. Thus, in the United States, participation in space launch activities is forbidden without the pertinent license issued by the American government. In this respect, the Act establishes that any person or entity, regardless of their nationality, must obtain a license to launch or reenter a space vehicle and to operate a launch or reentry site.[500] Additionally, all US citizens or legal entities, including entities

---

[495] 49 U.S.C. Subtitle IX, ch. 701--Commercial Space Launch Activities, Secs. 70101-70119, formerly the Commercial Space Launch Act of 1984 (CSLA), as amended (49 U.S.C. App. 2601-2623) [hereinafter CSLA].

[496] Before the enactment of the Commercial Space Launch Act, the industry faced absurd bureaucratic obstacles while procuring the authorization of space launches. For example, the 1982 request for the launch of Conestoga required the consent of several agencies, including the Bureau of Alcohol, Tobacco and Firearms. Thus, the Commercial Space Launch Act abrogated the authority of all agencies other than the Department of Transportation, with the exception of the Federal Communications Commission which requires authorization for the launch of communications satellites governed by the Communications Act of 1934 and Secretary of Commerce authorization for the activities governed by the Land Remote-Sensing Commercialization Act. Filiato, *supra* note 254 at 777.

[497] However, the US private sector only actually started to provide launch services on a regular basis only after the adoption of the 1988 Amendments and NASA's decision to cease competing with the Space Shuttle. "Space Risk Management", *supra* note 67 at 48.

[498] P. L. Meredith, "A Comparative Analysis of United States Domestic Licensing Regimes for Private Commercial Space Activities", (1989) 32 *IISL* at 374 [hereinafter "A Comparative Analysis of US Licensing Regimes"].

[499] In order to implement this system, the Federal Congress vested the Secretary of Transport with powers to act as the competent authority of the act.

[500] 49 USC Section 70104 (a) (1).

organized in foreign jurisdictions, which are controlled by US citizens or entities, require a license to operate launch vehicle outside the territory of the United States[501]. The Act also prescribes that entities organized or existing under the laws of a foreign country in which a US citizen or legal entity holds a controlling interest are to obtain a US license when launching or reentering a vehicle or operating a launch or reentry base in a foreign country, unless there is an agreement between the United States government and the foreign country where the foreign country has jurisdiction over the launch or reentry operation or said foreign country has jurisdiction by virtue of territoriality.[502] The Act thus covers launches made both outside the US and outside the territory of any foreign nation. The purpose of this provision is to prevent US entities from setting a corporation abroad and launching vehicles from the high seas, the inner space or even from Outer Space.[503]

4. 2. License procedure

The license procedure consists of four phases (i) submittal of application, (ii) safety review, (iii) mission review and (iv) issuance of license. The application, which must be filed with the Office of the Associate Administrator of Space Transportation, must contain: (i) the general requirements of applicant, (ii) information related to safety review, (iii) information related to mission review, and (iv) information related to payload determinations.[504]

The safety review which leads to the safety approval is, together with the mission review, the most important stage of the license procedure. Its purpose is to determine that the proposed launch will not endanger the public safety of the United States. Thus, four aspects are mainly examined: (i) the launch site, (ii) the quality procedures, (iii) personnel and (iv) the vehicle equipment.[505]

---

[501] 49 USC Section 70104 (a) (2).
[502] 49 USC Section 70104 (a) (4).
[503] 49 USC Section 70104 (a) (3).
[504] 49 USC Section 70105.
[505] 14 CFR § 415.

Mission review plays a residual role in the licensing procedure[506]. All aspects of the launch not covered in the safety review are examined in this stage. It aims at verifying that the launch abides by the national security and foreign policy interests and the international obligations of the United States. The regulations provide that mission approval is granted unless some element of the proposed launch poses a threat to US national security or foreign policy interest, constitutes a hazard to public health and safety or safety of property, or it is inconsistent with international obligations of the United States[507]. According to Kayser, this provision seems to put the burden of proof on the Office of Commercial Space Transportation and the applicant is not to demonstrate that its mission and the payload comply with all requirements.[508]

The most important element in the mission review is the examination of the payload[509]. The procedure differs whether the satellites are subject to a specific authorization regime, such as the ones foreseen for telecommunications and remote sensing, which are analyzed below, or whether the payload does not have such a regime[510]. With respect to the former, the Commercial Space Launch Act does not suppress or modify the authority granted to the Federal Communications Commission or National Oceanic and Atmospheric Administration[511]. Thus, the Office of Commercial Space Transportation requires that telecommunications satellites have obtained a license from the Federal Communications Commission before starting the mission review procedure. The same applies to remote sensing satellites, which must have obtained the license from the National Oceanic and Atmospheric Administration. For the rest of the space objects, including

---

[506]Information provided for this review should include the following: (i) the launch range, (ii) the number of launches planned and the targeted schedule, (iii) a general description of the launch vehicle, identifying the ELV contractors and manufacturers, (iv) a general description of the launch trajectory and ground track, including overflight of land masses and the sequence of major events from liftoff to payload impact for suborbital missions, or delivery on-orbit, (v) the orbital mission, (vi) the owner/operator of the payload, (vii) the payload function, (viii) whether the payload is US. Government owned, or licensed by another government agency, such as the FCC or NOAA, and (ix) any materials involved in this mission that could pose a unique hazard to the public National Transportation Safety Board, Special Investigation Report, NTSB/SIR-93/02, February 9, 1993.

[507]"Private Launch", *supra* note 30 at 99.

[508]*Ibid.* at 99.

[509] C. D. Williams, "Space: The Cluttered Frontier", (1995) 60 J. Air L. & Com. at 1171.

[510] W. B. Wirin, "Space Debris" (1989), 32 *IISL* at 189.

[511] 14 CFR 415.57.

foreign and US payloads there is a specific procedure for the obtainment of the mission review approval, where the Secretary has ample authority[512].

The issuance of the license constitutes the last phase of the license procedure[513]. Once all the previous steps have been satisfactorily fulfilled, the Secretary must verify compliance with the National Environmental Policy Act and examine the environmental impacts of the proposed launch. If the projected launch activity abides by the NEPA requirements, then the Secretary of Transportation will issue the license.[514] The denial, suspension, revocation and modification of a license are subject to administrative and judicial review[515].

4.3. Satellite telecommunications regulations

In the United States the authorization of telecommunications satellites[516] is governed by the Communications Act of 1934[517] and the implementing regulations[518] dictated by the Federal Communications Commission,[519] which is in charge of regulating all communications originated and received in the

[512] "A Comparative Analysis of US Licensing Regimes", *supra* note 498 at 374.
[513] 14 CFR 415.9
[514] 42 U.S.C. 4321, et seq. Applicants may be required to provide additional information concerning the environmental effects of a proposed launch activity when any of the following cases exists: (a) proposed new launch sites not covered by existing environmental documentation; (b) a proposed new launch vehicle with characteristics falling measurably outside the parameters of existing environmental documentation; (c) proposed launches from established sites involving vehicles with characteristics falling measurably outside the parameters of the existing environmental impact statement covering those sites; (d) A proposed payload that may have significant environmental impacts in the event of a launch accident; (e) other factors as determined by OCST.
[515] *Ibid.*
[516] The build, launch and operation of a satellite is subject to a construction permit, a launch authorization and an operating license.
[517] Communications Act of 1934, 47 USCA.
[518] 47 CFR 100.19 (direct broadcasting satellites), 47 CFR.391 (domestic fixed satellites), and 47CFR 25.392 (radiodetermination satellites).
[519] Satellites intended for international communications are also governed by the Communications Act of 1962. The Federal Communications Commission established rules under the Communications Act for the issuance of a license. In this respect, it established that "no person shall use or operate apparatus for the transmission of energy or communications or signals by space or earth stations except under, and in accordance with, an appropriate authorization granted by the Federal Communications Commission. 25 CFR 102.

United States, as well as, distributing and regulating frequency bands and orbital locations for communications satellites[520].

There are three types of license application possibilities. The first one and simplest is when an applicant seeks a license for a system where spectrum has already been allocated and rules already exist for the licensing of the system[521]. The second one concerns applicants seeking the establishment of a new system, which gives rise to questions of spectrum allocation and service rules. The last type concerns the application for a license where either service rules or a spectrum allocation need to be created[522].

Additionally, private commercial telecommunications satellite services have been classified in five different categories: (i) fixed satellite services, (ii) direct broadcast satellite services, (iii) radiodetermination satellite services, (iv) international fixed satellite system, also called separate system and (v) mobile satellite services.

The regulations prescribe that an applicant for a satellite services license must be legally, technically and financially qualified and it must file sufficient information with the FCC to evidence that it complies with the requirements to obtain a license[523]. Legal qualifications deal with the character and ownership of the applicant.[524] Financial qualifications refer to

---

[520] The frequency bands for fixed satellite services are listed in the Code of Federal Regulations and include separate bands for uplink and downlink.

[521] A satellite operating license allows the holder to use a specific orbital location and associated frequencies for a limited period of time. P. L. Meredith, "Implementing a Telecommunications Satellite Business Concept: Overview and Relative Timing of Legal Actions", (1990) 33 *IISL* at 43.

[522] G. V. D'Angelo, *Aerospace Business Law* (Westport, Conn.: Quorum Books, 1994) at 85.

[523] The regulations also spell out specific information that must be provided with each application so that the FCC might efficiently and effectively grant or deny the requests for licenses. N. C. Goldman, *American Space Law International and Domestic* (Ames: Iowa State University Press, 1988) at 177.

[524] In this respect, there are several restrictions to the ownership and control by foreigners of telecommunications services classified as common carriers. The Telecommunications Act of 1996 continues to limit ownership by foreigners of both stock and voting rights to 20 and 25 percent respectively as did the Federal Communications Act of 1934, but now permits foreigners to lawfully exercise control in their capacities of directors and officers. According to Gregory Sidak, it is simply irrational for Congress to say as a matter of federal telecommunications policy that active foreign managers may control a US radio licensee but passive foreign investors may not. However it bears adding that ownership restrictions do not refrain licensing

the applicant's ability to secure sufficient funds to proceed with the construction, launch and operation of the proposed satellite system. Requirements for financial qualifications vary from one category of satellite service to the next. Technical requirements concern the technical characteristics of the proposed satellite system, as well as the ability to operate it. The applicant must demonstrate in the satellite application that it meets the satellite design and performance requirements prescribed by the FCC for the proposed service and that it has or intends to hire technically qualified personnel. The FCC has always maintained a minimalist approach to technical regulation of satellite systems, giving the licensees maximum flexibility in the design of the system. Like the financial requirements, technical requisites vary according to the type of satellite services[525].

Any decision made by the FCC regarding a petition for a license involves the FCC's analysis of whether the proposed service complies with the public interest requirement. Indeed, the Commission has to evaluate whether the public interest, convenience, and necessity will be served by the award of the license. If the FCC arrives at a positive conclusion the license is to be granted[526].

The processing of an application differs according to the category and type of service[527]. In addition to the formalities of the application, and the verification of the legal, financial and technical requirements, the FCC also examines the following criteria: (i) whether the satellite service would cause interference to other users of the spectrum, (ii) whether it would create air hazard, and (iii) whether it would impose a radiation threat or violate environmental laws[528].

---

aliens from operating satellites on a private, non common carrier basis. J. Gregory Sidak, Foreign Investment in American Telecommunications, (Chicago: The University of Chicago Pres, 1997) at 135.

[525] "Case Study", *supra* note 65 at 125.

[526] J. Hermida, "Satellite Reciprocity Agreement between the United States and Argentina", (1998) 27 *J.Sp.L.* at 38 [hereinafter "Satellite Reciprocity"].

[527] In general it begins with the submission of the petition and the payment of fees. If the application is for domestic services it is processed in the Satellite Radio Branch of the Common Carrier Bureau's Domestic Facilities Division. Petitions of applicants seeking a license to provide international services are referred to the Office of Assistant Bureau Chief, International. When the proposed service is both international and domestic, the processing would normally begin in the domestic branch and then move to the international division.

[528] D'Angelo, *supra* note 523 at 87.

The most important recent development in the FCC licensing procedures was the use of auctions to decide who would receive the first set of personal communications service licenses. Through most of its history, the FCC assigned licenses on a first-come, first-served basis, through proceedings to determine which applicants would best serve the public interest or through lotteries. Only with the passage of the Omnibus Budget Reconciliation Act of 1993 did the FCC acquire the authority to sell licenses through an auction process, which brought billions of dollars to the US Treasury.[529]

4.4. Remote Sensing Satellites

Remote sensing satellites also have a specific set of rules which govern the conditions and requirements for their authorization[530]. In this respect, the Department of Commerce is to award licenses to private sector parties to operate private remote sensing space systems[531]. In this regard, no person subject to the jurisdiction or control of the United States, may directly or through any subsidiary or affiliate, operate any private remote-sensing space system without the appropriate license from the Department of Commerce[532].

The requirements to obtain such license are: (i) to operate the system so as to preserve the U.S. national security and to observe the international obligations of the United States, (ii) to provide the government of any country (including the United States) with unenhanced data concerning the territory under the jurisdiction of such government as soon as such data are available and on reasonable terms and conditions, (iii) to obtain and provide the government with specific unenhanced data, (iv) to make disposition of any satellites in space in a manner satisfactory to the president upon termination of the operations under the license, (v) to furnish with complete orbit and data collection characteristics of the system and to inform the Secretary of Commerce immediately of any deviation, and (vi) to notify the Secretary of Commerce of any agreement intended to be executed with a foreign nation, entity, or consortium involving foreign nations or entities[533].

The Secretary of Commerce must review any license application and make a determination within 120 days from the submission of such

---

[529] C. H. Kennedy & M. Veronica Pastor, *An Introduction to International Telecommunications Law* (Boston: Artech House, 1996) at 193.
[530] Michael R. Hoversten, "U.S. National Security and Government Regulation of Commercial Remote Sensing From Outer Space" (2001) 50 *A.F. L. Rev.* at 253.
[531] 15 CFR 960.
[532] 15 USC 5601.
[533] 15 CFR 960.1

application. Protection against the competition of existing licensees does not constitute valid grounds to deny a license application. The Secretary of Commerce may impose penalties for noncompliance with the license requirements or other provisions of the Act. The maximum penalty may not exceed $10,000[534]. Any final decision of the Secretary of Commerce, even the denial of a license, is subject to judicial review[535].

The common features shared by these three authorization regimes fashioned by the United States to implement the authorization obligations have been proposed to serve as a model for future licensing regimes in other countries[536]. However, as arises from the above discussion they also adopt standards, such as the national interest and US foreign policy, as legal grounds for the award of the license, which have been widely contested by international scholars[537] and whose legality in light of the international scenario is, at best, rather dubious.

4.5. Continuing supervision

In order to implement the continuing supervision principle the US has vested the Secretary of Transportation with powers to monitor activities of the licensees. For this purpose, the Secretary of Transportation may designate an officer as an observer at a launch site or reentry site which the licensee uses, at a contractor's production facility or assembly site used to produce or assemble a launch or reentry vehicle and at the site where the satellite or other payload is integrated with a launch or reentry vehicle. The observer must monitor the activity and must seek compliance with the license. The licensee is required to cooperate with said observer[538]. The Secretary of Transportation is entitled to conduct investigations and inquiries and to enter at a reasonable time a launch or reentry site, and a production facility, assembly or integration site and even to seize a satellite or space launch vehicle or any other object used in violation of the statutory provisions[539].

---

[534] Each day of operation in violation of the requirements and regulations constitutes a different infringement.

[535] "A Comparative Analysis of US Licensing Regimes", *supra* note 498 at 374.

[536] As summarized by Pamela Meredith, they designate a federal licensing agency, they impose licensing requirements, they set forth the criteria which must be satisfied for a license to be issued, they establish application and application review procedures and they provide for suspension or revocation of a license. "A Comparative Analysis of US Licensing Regimes", *supra* note 498 at 374.

[537] Straubel, *supra* note 510 at 30; "Private Launch", *supra* note 30 at 99.

[538] 49 USC Sec. 70106.

[539] 49 USC Sec. 70108.

The Secretary may suspend, revoke[540] or even prohibit[541] a license in the event that the licensee has not complied with its requirements and even if it is considered that the revocation or suspension is necessary to protect the public health and safety, the safety of property, or a national security or foreign policy interest of the United States. Similar faculties have been given to the Federal Communications Commission and the Department of Commerce.

5. IMPLEMENTATION OF THE REGISTRATION OBLIGATIONS

In order to assist the U.S. government in implementing Article IV of the Registration Convention[542], it is prescribed that each licensee of a launch service is responsible for registering all objects placed in space in the course of conducting activities authorized by its license, except for objects owned by a foreign entity or the US government[543]. Registration of objects owned by a foreign entity is the responsibility of that foreign entity.

The standards of article IV of the Registration Convention serve as the basis for the information to be provided by the licensee. Thus, according to US regulations, within 30 days after the launch, each licensee has to submit to the Office of Commercial Space Transportation the following data concerning any vehicle or other object it has launched into outer space: (i) the international designator of the space object(s); (ii) date and location of launch; (iii) final orbital parameters, including: (a) nodal period, (b) inclination, (c) apogee; and (iv) general function of the space object.

6. CONCLUDING REMARKS

As can be concluded from the above analysis of US domestic space law, from a legislative technique perspective, US national space law consists of a series of laws and regulations which govern specific aspects of different space activities, as well as several non specific norms which have a direct impact on the space industry. This reflects the US early involvement in the exploration and use of outer space, and its approach to deal with space law issues as its needs arose. As a consequence of this reactive approach, the US lacks a single or primary legislative instrument that governs all its national space activities. Nonetheless, US domestic space law is thoroughly comprehensive and covers

[540] 49 USC Sec. 70107.
[541] 49 USC Sec. 70108.
[542] Registration Convention, article IV.
[543] 14 CFR Sec. 415.81.

all possible areas of outer space exploration and use. Furthermore, a sophisticated system of regulations has been implemented through the actions of administrative agencies involved in the governance of space activities.

From a law reform standpoint and participatory theory, the US system of public notice for the adoption of regulations, which requires that federal regulations and other agency rules be promulgated after a public notice-and-comment procedure, allows ample participation of all those interested in the formulation of the regulations.

The United States have implemented a thoroughly comprehensive system to reallocate its liability attributed by the Outer Space Treaty and the Convention of Liability according to its space policy objectives. Under this system, first party risks are allocated through a system of mandatory reciprocal waivers of liability whereby each party agrees to be responsible for any damage or loss resulting from activities carried out under the use of a space license. Second party risks are distributed in a two-layered basis, where the private launch operator assumes the risk of losses through a system of insurance or self-insurance up to the amount of the maximum probable loss, i.e., $100,000,000 or the highest amount of liability insurance available, at a reasonable cost, on the world market in case the same is lower than $100,000,000, and the government absorbs the risks from that limit upwards through the so called waivers of liability. Third party risks are allocated between the private launch provider and the government on a horizontal basis, consisting of three layers. In the first layer, risks are absorbed by the private space launch provider through insurance or demonstration of financial responsibility in an amount sufficient to compensate the maximum probable loss, which has been capped at $ 500,000,000, or the maximum liability insurance available on the world insurance market at a reasonable cost. Risks in the second layer are assumed entirely by government indemnification up to the amount of $ 1,500,000,000. The third layer includes all claims above the upper limit of the indemnification and is the exclusive responsibility of the launch provider. This risk allocation system permits the United States government to redistribute the liability which the international space law instruments have attached to it to the different space players in accordance to its objectives of promoting the strength of its private sector space launch industry while safeguarding the public safety interests of the US population in general.

The US has structured the authorization requirement through a system of licenses for space launches and reentry operations, for the operation of launch and reentry sites, and for telecommunications and remote sensing satellites. The first two categories are dealt with in the Commercial Space

Launch Act, which mainly requires a US license for all launches in the US territory and for the activities of US citizens or corporations outside the United States. The most important aspects of the license procedure are the safety review and the mission review. This double review process allows the government to control the two areas which concern it most, i.e., public security and public health through the safety review, and national security and foreign policy interests through the mission review. However, the mission review may lead to situations which may contravene international law. Also, the competent authority in charge of the license procedure verifies compliance with the National Environmental Policy Act and examines the environmental impacts of the proposed launch.

The license of telecommunications satellites has been entrusted to the Federal Communications Commission and is governed by a specific set of rules which applies only to these satellites. There are different types of license application procedures and private commercial telecommunications satellite services have been classified in different categories. Basically, all applicants must be legally, technically and financially qualified and must satisfy the public interest test, which requires that the public interest, convenience, and necessity will be served by the award of the license.

Remote sensing satellites also have a specific set of rules which governs the conditions and requirements for their authorization. For this purpose, the Department of Commerce has been entrusted with authority to award licenses to private sector parties to operate private remote sensing space systems. The regulations prescribe that no person subject to the jurisdiction or control of the United States may operate any private remote sensing space system without the appropriate license from the Department of Commerce.

The United States has also implemented a system for the continuing supervision of the activities carried out by non governmental entities, whereby a governmental officer is appointed as observer at the premises of the non governmental entity. Furthermore, the government may suspend, revoke or even prohibit a license in the event that the licensee has not complied with its requirement or if these extreme measures are necessary to protect the public health and safety, the safety of property, or a national security or foreign policy interest of the United States.

Finally, the US adopted simple and straightforward measures to comply with the registration obligation arisen from the Outer Space Treaty and the Registration Convention. Basically, each licensee of a launch service is responsible for registering all objects placed in space in the course of conducting activities authorized by its license, except for objects owned by a foreign entity or the US government. Additionally, the licensee has to submit the information contained in article IV of the Registration Convention to the Office of Commercial Space Transportation.

B. RUSSIAN FEDERATION

1. NATIONAL APPLICATION OF INTERNATIONAL LAW

As head of the State, the President represents the Russian Federation in the conduct of international relations. In such role, the President has been entrusted with the direction and conduct of Russia's foreign policy and has been accorded primary treaty-making responsibility, which includes the negotiation and conclusion of international treaties, as well as the signature of ratification documents[544]. The Constitution prescribes that the subjects of the Russian Federation, i.e., republics and provinces, have the right to conduct their own international and foreign economic relations with foreign states. As argued by a contemporary Russian publicist, "this provision implies that the subjects of the Federation are granted limited treaty-making power, at least for matters over which they have exclusive jurisdiction."[545] The Russian constitution[546] has also reserved the ratification functions to the federal parliament.[547] However, the Russian Constitution does not appear to be clear with respect to which treaties are subject to ratification.[548]

After the collapse of the Soviet Union, the Russian Federation radically changed the long-standing Soviet dualist approach and expressly accorded full supremacy to international law[549]. Thus, according to the Russian Constitution of 1993: "The universally-recognized norms of

---

[544] Konstitutsiia (Rossiiskoy Federatsii) [Constitution of the Russian Federation] (1993), Rossiiskaia Gazeta, Dec. 25, 1993, article 80.4 [hereinafter "Russian Constitution"].

[545] G. M. Danilenko, "The New Russian Constitution and International Law" (1994) 88 *A.J.I.L.* at 453.

[546] "Russian Constitution", article 106.

[547] The Federal Council must consider those federal laws adopted by the State Duma relating to the ratification and denunciation of international treaties of the Russian Federation. Danilenko, *supra* note 545 at 454.

[548] *Ibid.* at 454.

[549] D. Michalchuk, "Filling a Legal Vacuum: The Form and Content of Russia's Future State Immunity Law Suggestions for Legislative Reform" (2001) 32 *Law & Pol'y Int'l Bus.* at 493.

international law and international treaties and agreements of the Russian Federation shall be a component part of its legal system. If an international treaty or agreement of the Russian Federation fixes other rules than those envisaged by law, the rules of the international agreement shall be applied"[550]. This clearly shows that international treaties override all conflicting domestic statutes. The Russian Federation has been enacting several commercial laws where disputes are meant to be decided in accordance with international law[551]. Additionally, the Constitution attributes competence to the Constitutional Court to hear cases dealing with international treaties, as well as, agreements which have not come into force. Since its inception, the Court has upheld international law over domestic statutes in several occasions[552].

2. NATURE OF RUSSIAN NATIONAL SPACE LAW

The Russian Federation has adopted a considerable number of norms to regulate outer space activities and to secure compliance with its obligations arising from the international Space Law instruments. The most important norm is the Law of the Russian Federation on Space Activity, which was adopted by the Supreme Soviet of the Russian Federation on August 20, 1993,[553] amended on October 4, 1996, and enacted on November 29, 1996.[554] Other norms include the Edict of the President of the Russian Federation about implementation of state policy in the field of rocket and space industry of 20 January 1998[555], which gives the Russian Space Agency responsibility to shape state technical policy in the industry, as well as, strategic missiles and military space programs, the decree of the government of the Russian Federation No. 533, 1 May 1996[556], which adopted the National Space Policy of the Russian Federation, the Decree of the Government of the Russian Federation No. 104 approving the Statute on the Licensing of Space Activity[557], the Decree of the Government of the Russian Federation No. 1282 about State support and backing for space activities,[558] which attributes

---

[550] "Russian Constitution", article 15.4.

[551] *Ibid.* at 494.

[552] Danilenko, *supra* note 545 at 454.

[553] Law of the Russian Federation about space activity, August 20, 1993 Resolution No. 5663-1.

[554] Federal Law about introduction of changes and amendments into the Law of the Russian Federation about space activity, adopted by the State Duma, October 4, 1996, enacted on November 29, 1996; No. 147-FZ.

[555] Edict of the President of the Russian Federation No. 54 of 20 January 1998.

[556] Decree of the government of the Russian Federation No. 533, 1 May 1996.

[557] Decree of the Government of the Russian Federation No. 104 about the approval of the the Statute on the Licensing of Space Activity, 2 February 1996.

[558] Decree of the Government of the Russian Federation No. 1282 about State support

special state importance to the Federal Space Program, and the Edict of the President of the Russian Federation[559], which established the Russian Space Agency, among many others.

The principal norm in the Russian Federation is the Law on Space Activity. It lays down the main legal scenario for the development of space activities, it establishes the organization of space activities in Russia and distributes responsibilities among different state entities. Furthermore, it sets the policy basis for future regulation and it mandates the adoption of a space policy instrument, called Federal Space Program of Russia[560]. As determined in its preamble, the purpose of the act is to "provide legal regulation for space activities and to stimulate the application of the potential of space science and industry for solving the socioeconomic, scientific, technical and defense tasks of the Russian Federation[561]".

The scope of the law is delimited by the concept of space activities under the jurisdiction of the Russian Federation. The concept of space activity is defined as "any activity immediately connected with operations to explore and use outer space...", and includes: space research; space communications, remote sensing of the Earth, satellite navigation, piloted space missions; manufacturing of products in outer space and any other kinds of activity performed with the aid of space technology[562]. This definition is of significant practical importance since the activities which are subject to authorization and continuing supervision are those which fall within this definition[563]. The law restricts the purpose of the space activities to the promotion of the well-being and security of the citizens of the Russian Federation, the development of Russian Federation, and the solution of global problems of mankind. Even if these goals appear to be considerably ample and far reaching in scope, this definition may curtail the development of space activities whose goals are not those expressly stated in the norm.

---

and backing for space activities, 11 December 1993.

559 Edict of the President of the Russian Federation about structure of management of space activity in the Russian Federation, February 25, 1992.

560 The Federal Space Program of Russia is considered as a document which sets the basis for the creation and use of space technology for scientific and national-economy purposes. Law of the Russian Federation on Space Activity, Article 8 (1).

561 Law of the Russian Federation on Space Activity.

562 Space activity comprises creating (including development, manufacture and test), as well as using and transferring of space techniques, space technology, other products and services necessary for carrying out space activity.

563 Law of the Russian Federation on Space Activity.

The law also sets the foundations for the organization of space activities. In this respect, it establishes federal competence for the regulation and administration of space activities in the Russian Federation. It also distributes the responsibilities among the different governmental bodies. In this respect, it confers jurisdiction to the Supreme Soviet of Russian Federation to determine the space policy of the Federation, which includes the adoption of legislative acts to regulate space activities, the adoption of the Federal Space Program of Russia, and its control, the ratification of international treaties of Russian Federation on issues of space activity, and the resolution of other issues, within its competence related to space activities[564]. The Supreme Soviet constitutes the highest level of the space management structure[565].

The law empowered the President of Russian Federation with responsibility for the implementation of space policy of Russian Federation and to ensure the day to day management of space activities[566], which includes the issuance of edicts and executive orders necessary for carrying out space activity, and the supervision of the activities of the Council of Ministers with respect to the implementation of the Federal Space Program[567]. The Council of Ministers must ensure the supervision of space activities, which encompasses the issuance of decrees and executive orders necessary for carrying out of space activities, the analysis of the draft Federal Space Program of Russia; the approval of Regulations of the Russian Space Agency; and the adoption of measures to protect interests of Russian Federation and its organizations and citizens in the field of space activity[568]. In turn, the Russian Space Agency must elaborate a draft of the Federal Space Program of Russia, [569] issue licenses for space activities, organize the certification of space technology, adopt the necessary normative technical documentation, and ensure the safety of space activity[570].

The Law affords little participation of interested individuals and entities in the elaboration of norms and regulations for the governance of

---

[564] Law of the Russian Federation on Space Activity, Article 5 (2).
[565] V. Postyshev & I. Moiseyev, "Space Policy in Russia. Perspectives for Legal Development", (1992) 35 *IISL* at 113.
[566] *Ibid.* at 113.
[567] Law of the Russian Federation on Space Activity, Article 5 (3).
[568] Law of the Russian Federation on Space Activity, Article 5 (4).
[569] This should be done in coordination with the Ministry of Defense, the Russian Federation, the Russian Academy of Sciences and other state customers of works in creation and use of space technology.
[570] Law of the Russian Federation on Space Activity, Article 6 (2).

space exploration and use. This participation is practically restricted to the possibility of participating in a competition for the elaboration of the Federal Space Program[571].

3. RESPONSIBILITY AND LIABILITY

The law contains several provisions regarding the reallocation of liability arising from space activities. These provisions are rather vague and they do not contemplate all possible scenarios for the redistribution of liability. Furthermore, the law refers many aspects to regulations, which have not been enacted so far.

3.1. First party risks

The law is completely silent with respect to first party risks. However, as arises from our previous findings if the launch carrier may be considered a US national under the extraterritorial effects of the Commercial Space Launch Act, it will have to abide by the first party risk regulations contained in US law[572].

3.2. Second party risks

The Law is also silent with regard to second party risks. Therefore, the general principle consecrated in the law, i.e., the organizations and citizens responsible for operation of the space technology involved must pay compensation for damage inflicted as a result of a space activity related accident will apply to damage caused to state property. However, it may be construed from the prescriptions of article 29 that this principle will not apply when the damage is the result of a violation of the law or regulations committed by a state body and its officials[573]. In such case, the liability will be diminished or eliminated according to the degree of the incidence of the illicit conduct of the state officials or their agencies.

---

[571] The procedure and terms for holding competitions of space projects for scientific and national-economy purposes was entrusted to the Russian Space Agency and the Russian Academy of Sciences was given a minor role in the elaboration of these procedures. Law of the Russian Federation on Space Activity, Article 8 (3).

[572] See *supra* note 460.

[573] Law of the Russian Federation on Space Activity, Article 29.

3.3. Third party risks

Damage to third parties is reallocated to the responsible, whether an organization or citizen, for operation of the space technology involved. The term responsible is not defined in the law or the regulations. However, an interpretation of this provision in conjunction with the norms dealing with the licensing systems indicates that the responsible for the operations may be equated with the licensee[574].

There is, however, one exception in the event that the damage is the result of errors committed at the creation and use of space technology, in which case liability for damage is imputed partly or fully on the appropriate organizations and citizens. The concept of organization and citizen that created or used space technology is ambiguous and imprecise[575]. The responsible, i.e., the licensee, undoubtedly uses space technology and so do many entities and individuals. Nonetheless, on account of the spirit of the law this provision may only be interpreted as meaning that in the event that an entity or individual other than the licensee caused or contributed to cause damage, that entity or individual will be wholly or partially liable. In the latter case, as well as in all other cases of joint liability, the injured party may claim compensation from all or any of the liable parties[576]. The organizations or citizens that paid for all damages to the victim may seek reimbursement of the amounts which exceeded their corresponding compensation from all other joint liable entities[577]. The basis of liability for damage inflicted by a space object of the Russian Federation within the Russian territory or outside the jurisdiction of any state, except outer space, is objective, i.e., it arises regardless of fault.[578]

The liability of the responsible entity, or the individuals or entities that created or used the space technology causing damage, is limited to the amount of the insured sum or insurance indemnity provided in the insurance contracts[579]. Liability exceeding the limit of the insurance may either still be the responsibility of the responsible entity - or the individuals or entities that

---

[574] *Ibid.*
[575] *Ibid.*
[576] *Ibid.* article 30 (3).
[577] *Ibid.* article 30 (3).
[578] *Ibid.* article 30 (3). If in any place, apart from the Earth surface, damage has been inflicted on a space object of Russian Federation or on property on board of such object by another space object, the liability of organizations and citizens must emerge with their being at fault and in proportion to their fault.
[579] Law of the Russian Federation on Space Activity, Article 30 (4).

created or used the space technology causing damage- through recourse against their property if so specified in the legislation of Russian Federation or it may be assumed by the Russian government on account of the guarantee undertaken under the Law. With respect to the latter case, "the Russian Federation shall guarantee full compensation for direct damage inflicted as a result of accidents while carrying out space activity in accordance with legislation of Russian Federation.[580]" The legislation could, of course, provide recourse up to a further limit and above that new limit the government guarantee would apply.

4. IMPLEMENTATION OF THE AUTHORIZATION AND SUPERVISION PRINCIPLE

The domestic implementation of the authorization principle in the Russian Federation is twofold. First, the Russian government adopted a licensing system for space activities, and second it established a system of certification of space technology.

4.1. Licensing system

The Law established a licensing procedure for the pursuit of space activities for organizations and citizens of the Russian Federation, as well as, for foreign organizations and citizens under its jurisdiction. The law makes it explicit that the space activities which are subject to the licensing system include tests, manufacture, storage, preparation for launching and launching of space objects, as well as control over space flights[581]. As arises from our examination of the international system, these activities, which are referred to as preparatory, related and concluding activities, do not require authorization under international law.[582] However, States are free to subject these activities to authorization.

---

[580] *Ibid.* article 30 (1).

[581] Space operations subject to licensing include the making (encompassing production and testing) of space-rocket complexes and their component parts, storage, readying for launch, launching and utilizing space vehicles, and also the control of space missions. Law of the Russian Federation on Space Activity, Article 9 (2). Decree of the Government of the Russian Federation No. 104 about the approval of the Statute on the Licensing of Space Activity, 2 February 1996, article 3.

[582] See *supra* Chapter I.

Under the Russian Federation legislation, the carrying out of space activities without a license is considered an offence, as well as carrying out space activities in willful violation of the terms of the license[583].

The types, forms, and terms of licenses, the conditions and procedures for their issue, withholding, suspension or termination, as well as other questions of licensing are regulated in the Statute on the Licensing of Space Activity, which was approved by Decree No. 104.[584]

Another standard which any space activity must comply with is safety[585]. This requisite is legislated with more precision than the other grounds for the award of the license.[586] In this respect, the law establishes with clarity that any space activity must be carried out with the observance of the safety requirements laid down by the legislation of Russian Federation.[587] This implies that all participants in a space activity are obliged to take all necessary measures to ensure that it is carried out without imposing any threat to public safety or the environment[588].

The Russian Space Agency is in charge of conducting the licensing procedure and awarding the license. It has been entrusted with ample faculties to carry out its responsibilities. However, in certain cases it must seek the advice of a commission of experts[589]. These cases encompass those space operations which are not included in Russian Federal Space Program and all commercial space projects[590]. The conclusions of the expert commission are not binding but in case they differ with the opinion of the Russian Space Agency the definitive decision is escalated to the chief executive officer of the Agency[591]. All determinations of the Space Agency are subject to judicial review before the courts or arbitration tribunals[592].

---

[583] Law of the Russian Federation on Space Activity, Article 9 (5).
[584] Decree of the Government of the Russian Federation No. 104 about the approval of the the Statute on the Licensing of Space Activity, 2 February 1996.
[585] Law of the Russian Federation on Space Activity, Article 22.
[586] *Ibid.*
[587] Overall guidance of the work to ensure the safety of space activity rests with the Russian Space Agency and the Ministry of Defense of Russian Federation.
[588] Law of the Russian Federation on Space Activity, Article 22.
[589] *Ibid.* article 11 (1).
[590] Decree of the Government of the Russian Federation No. 104 about the approval of the Statute on the Licensing of Space Activity, 2 February 1996, article 18.
[591] Law of the Russian Federation on Space Activity, Article 11 (4).
[592] *Ibid.* article 9 (5).

4.2. Certification of Space Technology

Space technology is subject to a certification system which ensures the compliance with the requirements established by the legislation of the Russian Federation through an inspection procedure[593]. Space technology which must be certified under this procedure includes space objects, ground and other objects of space infrastructure created for scientific and national economy purposes, and equipment used in the creation and use of space technology[594].

The certification procedure is also conducted by the Russian Space Agency, which must award a certificate to each sample of space technology that complies with the terms and procedures contained in the regulations[595].

According to the report prepared by COPUOS on the review of the existing national space legislation, despite the silence of the law and its regulations, the certification procedure is implicitly directed to non governmental entities only[596].

4.3. Continuing supervision

In order to comply with the continuing supervision principle, the Russian Space Agency has been entrusted with power to monitor the licensed operations, as well as, to query the licensee about its observance of the license conditions and with the faculty to suspend or annul a license whose holder does not abide by the applicable conditions or regulations[597]. Furthermore, the Agency may even shut down the operations of readying for launch, or other operations at the site where space operations are conducting if they threaten public health and safety or the state interests and security of the Russian Federation. The Agency is also entitled to modify the license in the event that the Russian Federation assumes new international obligations[598].

Additionally, the licensee must allow persons designated by the Russian Space Agency to verify the conditions of use of the license and must satisfactorily answer all queries of the Russian Space Agency relative to the licensed operations.

---

[593] *Ibid.* article 10 (1).
[594] *Ibid.* article 10 (1).
[595] *Ibid.* article 10 (1).
[596] A/AC.105/C.2/L.224 at 7.
[597] Decree of the Government of the Russian Federation No. 104 about the approval of the Statute on the Licensing of Space Activity, 2 February 1996, article 22.
[598] *Ibid.* article 25.

5. IMPLEMENTATION OF THE REGISTRATION OBLIGATIONS

The Law on Space Activities prescribes that all space objects of the Russian Federation must be registered and must bear markings certifying their appurtenance to Russian Federation[599]. The law does not specify the information that has to be provided so that the Russian government can comply with its obligations under article IV of the Registration Convention[600].

In accordance with the provisions of the Outer Space Treaty and the Registration Convention, the Russian law clarifies that the Russian Federation retains jurisdiction and control over space objects registered in it during their ground time of such objects, at any stage of a space flight or in outer space, and also on return to the Earth outside the jurisdiction of any state.[601] The same applies to crew of a piloted space object registered in the Russian Federation[602].

6. CONCLUDING REMARKS

From a legislative technique perspective, the Act meets most of the characteristics of a framework law as it is wide-ranging in nature, it provides a general regulatory scenario, it sets the policy basis for future regulation and it refers the regulation of specific aspects of space activities to other agencies[603]. Furthermore, it lays down the organization of space activities and distributes responsibilities among different governmental entities, reserving ample regulatory and controlling powers to the legislative body. Compared with other national space acts[604], the Russian Law is evidently more comprehensive on account of Russian's vast participation in space activities. Its drafting technique, however, is most of the times imprecise and ambiguous and its text permits multiple interpretations which could have been avoided had the legislator employed a more straightforward approach.

From a law reform standpoint and participatory theory, the Russian Law gives marginal participation, if any, to interested individuals and entities

---

599 Law of the Russian Federation on Space Activity, Article 17 (1).
600 *Ibid.*
601 Law of the Russian Federation on Space Activity, Article 17 (2).
602 *Ibid.* article 20 (4).
603 Nolon, *supra* note 48 at 685.
604 See Australian and the United Kingdom acts below.

in the elaboration of norms and regulations for the exploration and use of outer space.

The law contains a risk distribution regime, which reallocates third party liability to the responsible entity, i.e., licensee, and the entity which created or used the space technology, if different from the licensee, in the event that said entity caused the damage. Liability is capped at the amount of the insurance and the government may assume part of the risks exceeding the insurance if so provided for by law.

The domestic implementation of the authorization principle in the Russian Federation is twofold. First, the Russian government adopted a licensing system for space activities, and second it established a system of certification of space technology. The legal grounds for the award of a license are drafted in vague and imprecise language, except for the public safety and the environment requirements. The Russian Space Agency has been entrusted with sufficient powers to continually supervise national space activities.

Although the Law prescribes that all space objects of the Russian Federation must be registered it does not specify the information that has to be provided for the fulfillment of Russia's obligations under the Outer Space Treaty and the Registration Convention.

C. AUSTRALIA

1. NATIONAL APPLICATION OF INTERNATIONAL LAW

Under Australian law, the federal executive, through the Crown's representative has exclusive and unfettered treaty-making power[605]. The federal power, nominally vested in the Queen, is exercisable by the Governor-General and, in practice, treaty making is carried out by the Federal Executive Council. Therefore, the Federal Executive Branch negotiates, concludes and even ratifies an international treaty. The role of the Australian Parliament is very limited. However, like in the United Kingdom, the government follows the practice of tabling international treaties before the two Houses of Parliament for a period of twelve days so that members can give any comment on the treaties before their ratification[606].

[605] N. D. Campbell, "Australian Treaty Practice and Procedure" in K. W. Ryan, ed., *International Law in Australia* (Sydney: The Law Book Company, 1984) at 53.
[606] *Ibid.* at 54.

Following the United Kingdom transformationist approach, international treaties become part of Australian domestic law only after they have been implemented by domestic legislation[607]. However, treaties relating to the conduct of war or treaties of cession can have direct application without the enactment of a domestic statute[608]. Additionally, the courts have held that when the Australian Parliament adopts national legislation for the express purpose of giving effect to a treaty, any conflict between a domestic statute and an international treaty must be resolved in the manner most consistent with the treaty[609]. With respect to customary international law, despite the important English precedents "the position in Australia is less clear and the few cases that have attempted to invoke customary international law have not generally succeeded"[610].

2. NATURE OF AUSTRALIAN NATIONAL SPACE LAW

Australia enacted a quite comprehensive domestic space law framework, which became effective in December 1998. As arises from the Explanatory Memorandum accompanying the Bill introduced to the Australian Parliament, the purpose of the Space Activities Act is to permit Australia to attract foreign companies to set launch facilities in its territory and waters, while meeting its obligations assumed under the International *Corpus Juris Spatialis*[611].

Unlike the United States, Russia or France, Australia does not have a space launch carrier industry. Therefore, its domestic regulatory framework has been conceived to encourage foreign companies to establish space launch facilities in Australia and its territorial waters. Thus, the whole system embodied in the Australian Space Activities Act tends to provide a favorable scenario for launch facility operators, while at the same time it intends to comply with the obligations assumed by Australia at the international level[612]. As stated in the Explanatory Memorandum: "the Government's wish to instill in prospective overseas investors and customers absolute confidence would be

---

[607] A. D Mitchell, "Genocide, Human Rights Implementation and the Relationship between International and Domestic Law: Nulyarimma V Thompson" (2000) 24 Melbourne U. L.R. at 16.
[608] *Ibid.* at 29.
[609] Mitchell, *supra* note 607 at 29.
[610] R. Balkin, "International Law and Australian Federalism" (1998) 92 *A.J.I.L.* at 793.
[611] Australian Space Activities Bill 1998, Explanatory Memorandum at 1.
[612] Australia is a signatory of the five international space law treaties.

undertaken in a certain legal and administrative environment consistent with international law and practice[613]."

The Act is the result of extensive consultation undertaken by the Department of Industry, Science and Resources with local interest groups and with major US and other foreign space industry players interested in providing services in Australia. It is far reaching in nature and since it is the first Act to deal with the regulation of space matters it does not need the amendment of other Australian legislation[614].

The Australian government established a simple administrative structure to implement the Act. For this purpose, the government created a licensing authority named Space Licensing and Safety Office, which provides a one-stop facility within government to deal with all aspects of launch authorization, except for accident investigation[615].

The Act comprises five substantive parts dealing with the regulation of space activities, i.e., licensing, liability for damage by space objects, registration of space objects, and investigation of accidents. The core of the norm is the licensing of space activities and the implementation of a risk distribution system for some of these activities. It also established a civil penalty regime for contraventions to the major obligations contained in the Act and a system of investigation of accidents and incidents modeled after the Chicago Convention provisions for the investigation of aircraft accidents. As the Chicago model, the system established by the Australian Parliament aims at preventing the occurrence of new accidents and incidents and not to assign blame or liability to any one[616].

Another significant feature of the Act is the fact that it "incorporates a high degree of flexibility in the regulations associated with the approval provisions [required under the Act]" by referring many regulatory aspects to future regulations, thus creating a reliable scenario within which evolutionary change can occur[617]. It also affords all interest groups the opportunity to negotiate the elaboration of this framework Act whenever a single project is proposed. For this purpose, the Explanatory Memorandum foresees the

613 Australian Space Activities Bill 1998, Explanatory Memorandum at 3.
614 *Ibid.* at 1.
615 *Ibid.* at 5.
616 Australian Space Activities Act 1998: No. 123, 1998, Part 7 Division 2, 87.
617 Australian Space Activities Bill 1998, Explanatory Memorandum at 5.

possibility of periodic reviews where the input of all interest groups will be an important contribution to any post-implementation review[618].

3. RESPONSIBILITY AND LIABILITY

The Australian Act on Space Activities contains a detailed and exhaustive regime on responsibility and liability. It sets forth the general principles of national responsibility and it contains a fairly comprehensive treatment of most cases of potential liability for damage caused by a space object.

The Act preempts any other legislation, norms and decisions regarding liability and thus compensation for damage caused by space objects to third parties is exclusively regulated in the Act. However, it does not prevent Australia from complying with any obligation to pay compensation under the Liability Convention or under international law[619].

The Australian Act attaches national liability for any damage the space object causes to a third party to the responsible party for the launch or return of a space object when damage occurs on Earth or to an aircraft in flight. In accordance with the international model, at the national level the Australian Act also exempts the responsible party of liability in the cases of gross negligence of the third party; or any conduct that the third party engaged in with intent to cause the damage. Additionally, the Act assigns liability to the responsible party for the launch or return of a space object in the event of damage to a space object launched or operated by a third party; or to a third party, or the property of a third party, on board such a space object, provided that the damage is due to the fault of the responsible party or of a related party. However, the Australian law limits the amount of compensation as examined in detail below in the case of a launch or return of a space object authorized by a launch permit or by an overseas launch certificate, provided that the damage did not result from a breach of any of the conditions of the permit, space license or certificate or the willful misconduct of the responsible party.

The Act identifies as the responsible party for the launch or return of the space object the following individuals or entities: (a) the holder of the permit in the case of a launch or return authorized by a launch permit; (b) the holder of the permission in the case of a return authorized by a permission; or (c) another person in the case of a return authorized by an agreement between the Minister and that other person. In the case of a launch or return that is not

[618] *Ibid.* 5.

[619] "Commercial Space", *supra* note 7 at 71.

authorized as mentioned in the preceding cases, but which is covered by an exemption certificate, the holder of the exemption certificate; or in the case of a launch authorized by an overseas launch certificate the holder of the certificate.

Also the following persons, provided that they are also Australian nationals, are considered responsible parties: (i) the person or persons who carried out the launch or return of the space object; (ii) any person who, at any time during the liability period for the launch or return, owned all or some of any payload forming part of the space object concerned; (iii) any other person specified in regulations made for the purposes of this definition.

Below we will analyze the way the Australian Act allocates risks and assigns liability to the different space players.

3.1. First party risks

In accordance with the objective sought by the legislator, the Space Activities Act does not contain any provisions dealing with the allocation of risks between the space launch carrier and its customer. It does, however, contain a general authorization so that the implementing authority may make regulations in relation to the waiver of some or all of the rights of persons connected with a launch, including their employees, contractors and subcontractors.[620]

Therefore, in principle and absent the introduction of specific regulations, the customer is free to negotiate with the launch carrier any scheme to distribute these risks. However, in our opinion, the negotiating power of the launch services provider and the characteristics of the launch market will in practice give little room for the customer to obtain a risk management approach substantially different from the ones used, for example, in the US private sector or by Arianespace.

Nonetheless, it bears noting that the above considerations apply to customers of space launch services providers which do not qualify as US nationals according to the Commercial Space Launch Act[621]. In effect, if the launch carrier may be considered a US national, it will have to abide by the first party risk regulations contained in US law. Also, if the launch carrier is a national of another state which has similar extraterritorial regulations, the customer of that carrier will be constrained to the first party risk allocation scheme contained in the legislation of its state of nationality.

---

[620] Australian Space Activities Act 1998: No. 123, 1998, Part 4 Division 1, 65.
[621] CSLA.

Thus, even if Australian law is silent with respect to the distribution of risks and the assignment of liability between the launch carrier and its customer, the risk management approach for the customer, in practice, will not deviate radically from the ones followed in other jurisdictions.

3.2. Second party risks

The Australian Space Act deals exclusively with the so called International Liability Risks, for the Commonwealth does not directly own or operate launch facilities and related range services[622]. Therefore, the legislator has been concerned mainly with establishing a risk allocation regime for the liability which Australia may face as a result of its obligations assumed under International Law. For this purpose, the Act established a two-tiered risk distribution system.

3.2.1. First layer: Insurance or Financial requirements

For a launch or return authorized by an Australian launch permit, as well as for a return authorized under section 43[623], the holder of the permit must insure the Commonwealth against any liability that it might incur under international law. The total insurance for each launch or return must be the amount of the maximum probable loss of damage to third parties caused by the launch or return.

Like the US Commercial Space Launch Act Amendments of 1988, the Australian Space Act adopted the maximum probable loss standard as the cap for the first layer of second party risks. However, unlike the US regime, the Australian system does not expressly foresee a monetary cap to the maximum probable loss standard. Nor does it take into account the amount of liability insurance available in the world market. Nonetheless, according to Michael Davis, Australia provides less demanding insurance requirements than the US. Davis predicts that Australian authorities will use the same MPL assessment methods, but since Australian territory is significantly less inhabited than the US, any MPL determinations in Australia will probably result in a considerably lower amount[624]. It is also worthy of note that the Act has especially authorized the possibility of introducing through regulations a different method for determining a minimum amount for insurance purposes.

---

[622] "Space Risk Management", *supra* note 67 at 32.
[623] Australian Space Activities Act 1998: No. 123, 1998, 43.
[624] M. E. Davis & R. J. Lee, "Financial Responsibility and Government Indemnities for Commercial Space Launch Activities - The Australian Approach", (1999) 50 *IISL* at 5 [hereinafter "The Australian Approach"].

Instead of obtaining insurance, the holder of the permit may opt to show direct financial responsibility for the launch or return. This possibility, modeled after the US Commercial Space Launch Act Amendments of 1988, allows the permit-holder to self-insure against the risks or resort to other risk management strategies, such as group risk retention or the acquisition of bonds, among many other alternatives[625].

The liability period for each launch of a space object is the period of 30 days beginning when the launch takes place[626]. The language used in the Act may trigger certain controversies, for it is sometimes difficult to determine when a launch takes place. However, it is expected that the regulations will provide a precise definition of the commencement of the liability period. This period is the same for carriers that acquire insurance and for those that opt to self insurance.

3.2.2. Second layer: Government assumption

The Australian government assumes all damages exceeding the amount of the insurance or financial responsibility which the launch operator must obtain or demonstrate[627]. There is no limitation for the assumption of these risks by the government so Australia will be assuming all risks above the maximum probable loss threshold. The Australian Act does not foresee the cases of exclusions in insurance policies. So in principle, the government may never assume damages which are below the maximum probable loss determination, even if it is not actually covered by the insurance policy obtained by the launch operator.

3.3. Third party risks

Third party risks are also allocated between the launch operator and the government on a horizontal basis, consisting of two layers.

3.3. 1. First layer: Insurance or Financial requirements

The provisions governing insurance requirements for second party risks also apply to third party liability risks. Thus, the launch operator must acquire liability insurance to protect against any compensation for damage to third

---

[625] J. Trieschmann, R. Gustavson & G. Sandra, *Risk Management & Insurance*, 9th ed. (Cincinatti: South Western College Publishing, 1995) at 4.
[626] Australian Space Activities Act 1998: No. 123, 1998.
[627] *Ibid.*

parties that the launch may cause. The total insurance for each launch or return must be the amount of the maximum probable loss of damage to third parties caused by the launch or return, which is governed by the same provisions applying to second party risks. The liability period is also the same as in the case of government risks.

The launch operator may acquire a separate policy from the one protecting the Commonwealth or may take out a single policy that insures itself against third party liability and the Commonwealth for second party risks[628]. Additionally, the launch operator may opt to show direct financial responsibility instead of hiring third party insurance[629].

Consistent with the international Space Law treaties and convention, the Australian Space Act establishes that the responsible party for the launch, defined as the holder of a permit for launches authorized by Australian authorities, is liable to pay compensation for any damage the space object causes to a third party on Earth or as a result of damage to aircraft in flight. The Australian Act also exonerates liability in the case of gross negligence of the third party and when the conduct that the third party engaged in was with the intent to cause damage. With respect to damage to other space objects, the responsible party is liable to the extent that the damage is due to its fault. The Australian Act also contains provisions establishing federal jurisdiction for actions dealing with liability and establishes statute of limitations compatible with those adopted by the Liability Convention. However, it neglected to consider the cases of standard exclusions in insurance policies which the US Commercial Space Launch Act does[630].

3.3.2. Second layer: Government assumption

The second layer of third party liability risks closely follows the model of the second party risks. The Australian government assumes all damages beyond the amount of the insurance or financial responsibility which the launch operator must obtain or demonstrate. Unlike the regime established under the United States Commercial Space Launch Act[631], there is no cap to the government's undertaking to assume third party risks and thus the Australian government even absorbs the maximum possible losses. However, in the case that the damage results from a breach of any of the conditions to the launch permit or the relevant space license the government's assumption of the third

---

[628] *Ibid.*
[629] *Ibid.*
[630] CSLA.
[631] CSLA.

party liability risks does not operate and the launch operator is solely responsible against third parties[632]. The same applies when the damage arises from a conduct engaged in by the launch operator with intent to cause damage and from its gross negligence[633].

4. IMPLEMENTATION OF THE AUTHORIZATION AND SUPERVISION PRINCIPLE

4.1. Licensing regime

The Australian Space Activities Act instrumented a two-level authorization approach consisting of licenses and permits, which has been envisioned to compete with a more burdensome US licensing system. In effect, the Australian authorities analyze the technical and safety considerations only once at the license level. Thus, the analysis for the launch permit, i.e., the authorization to actually carry out the launch, basically deals with insurance requirements, trajectory and type of payload. As arises from our above examination, the US system requires the verification of all the requirements for the issuance of the license in each launch or launch series.[634]

4.2. Space Licenses

The Act prescribes that all persons require a space license to operate a launch facility in Australia and for each kind of launch vehicle proposed to be used. The Act also foresees the scenario where the operation of the launch facility or the launch vehicle has been licensed by a foreign state, in which case the foreign license holders must seek an exemption certificate from Australian authorities. Australian nationals engaged in launch operations abroad are also subject to the obtainment of Australian authorization[635].

A company or individual seeking a space license to operate a launch facility or for a particular kind of space vehicle must demonstrate the following to the Australian authorities: (i) competence to operate the launch facility and the launch vehicle, (ii) compliance with environmental approvals, (iii) financial capacity to construct and operate the launch facility, (iv) the improbability of substantial damage to the public health, public safety and property, (v) non interference with Australia's national security, foreign policy

---

632 Australian Space Activities Act 1998: No. 123, 1998.

633 *Ibid.*

634 "The Australian Approach", *supra* note 624 at 3.

635 Australian Space Activities Act 1998: No. 123, 1998, Parts 11, 13 and 15.

or international obligations, and (vi) compliance with the criteria, if any, prescribed in the regulations[636].

Space licenses are issued for a specific period, which may extend up to twenty years, a feature unseen in other launch licensing systems[637].

4.3. Return of space objects

The Space Activities Act also establishes the need to seek authorization for the return of both national and foreign space objects. The former need either a launch permit or an exemption certificate whereas the latter requires authorization[638].

4.4. Launch permits

After the obtainment of the space license for the launch facility and the launch vehicles the next step is the issuance of a space permit for a particular launch. The Act establishes that Australian individuals and entities engaged in space activities in Australia are required to seek a launch permit from the space authorities both for launch operations within Australia and abroad. Foreign entities and individuals are to obtain a launch permit for launch activities within Australia[639].

Launch permits may be issued if the person seeking the permit (i) holds a space license, (ii) is a corporation, (iii) is competent, (iv) has satisfied all the insurance/financial requirements established under the Act, and (v) complies with all the criteria prescribed by the regulations. For the issuance of the permit, competent Australian authorities must be satisfied that the probability that the launch may cause substantial damage to public health, safety or property is low, and that the space object does not contain a nuclear weapon or other weapon of mass destruction[640].

4.5. Continuing Supervision

The Act also contains a regime aimed at implementing the continuing supervision obligations assumed by Australia under the Outer Space Treaty. Thus, for each licensed launch facility the Act requires the Minister to appoint

---

[636] *Ibid.* Division 2 Part 18.
[637] *Ibid.* Division 2 Part 19.
[638] *Ibid.* Division 1 Parts 13 and 14.
[639] *Ibid.* Division 3 Part 26.
[640] *Ibid.* 1998, Division 3 Part 26.

a launch safety officer, who has to ensure that any required notice is given of launches conducted at the launch facility and that no person or property is endangered by any launch until the space object is safely in Earth orbit or beyond. The launch officer must also monitor the compliance of space license or launch permit holders with the conditions of the license or permit[641].

To exercise these functions the launch safety officer has been entrusted with ample powers. In this respect, he or she may do all that is reasonably necessary or convenient for the performance of these functions. In particular, the launch officer may: (a) with the consent of the holder of the relevant space license enter and inspect the facility and any space object at the facility; and inspect and test any other equipment; (b) ask the licensee or permit holder to give him or her any necessary information or assistance, and (c) give any directions about the launch of a space object carried out, or proposed to be carried out, at the facility that he or she considers necessary to avoid any danger to public health or to persons or property. This includes the possibility to order the licensee or the permit holder to stop the launch or to destroy the space object, both before and after it is launched[642]. Additionally, the license may be suspended if the holder contravenes a license condition or the suspension is necessary for Australia's national security, foreign policy or international obligations[643].

The Act refers the establishment of a procedure for giving and complying with directions to the regulations. Lack of compliance with the directions of the launch safety officer is considered a criminal offence, which is sanctioned with a penalty of up to 100 units[644]. However, the Act makes it clear that the launch safety officer is not entitled to be involved in the normal business operations of the holder of a space license or launch permit[645].

5. IMPLEMENTATION OF THE REGISTRATION OBLIGATIONS

The Australian Act prescribes that the Minister for Industry, Science and Resources must keep a Register of Space Objects.[646] The following information must be recorded with respect to any space object launched under an authorization provided under the Australian Act: (i) the registration

---

[641] *Ibid.* Division 8 Part 50.
[642] *Ibid.* Division 8 Part 51.
[643] *Ibid.* Part 3 Division 2, 25.
[644] *Ibid.* Division 8 Part 53.
[645] *Ibid.* Division 8 Part 52.
[646] *Ibid.* Part 5 Division 76.

number given by the Minister for Industry, Science and Resources to the space object when it is granted a launch permit must, (ii) the launch facility, (iii) the date of the launch, (iv) basic orbital parameters, including: (a) nodal period, (b) inclination, (c) apogee and perigee, (v) the space object's general functions, (vii) the name of other launching states, and (vi) any prescribed particulars[647].

As arises from the foregoing enumeration, the data prescribed by the Australian Act follows, in general, the guidelines identified in the Registration Convention. However, in consonance with its nature, the framework Act foresees the possibility of requesting additional information.

6. CONCLUDING REMARKS

As arises from the foregoing analysis, from a legislative technique perspective, the Act meets all the characteristics of a framework law[648]. In effect, it is comprehensive in nature, it intends to provide a general regulatory scenario without legislating in detail every single aspect of the space industry and it provides solutions for the development of non-regulated areas through the delegation of authority to the Governor-General to make regulations on matters required or permitted by the Act or necessary or convenient for the development of the Act. From a law reform perspective, the Act is the result of extensive negotiations and consultations with local and international interest groups and Australian authorities committed to engaging in new rounds of consultation for any post implementation legislative review[649].

The Act preempts any other legislation, norms and decisions regarding liability and thus compensation for damage caused by space objects to third parties is exclusively regulated in the Act. It devised a simple licensing system which does not impose excessive burdens to the space players, but which ensures that the risk of damage is low. It also implemented a detailed and exhaustive regime on responsibility and liability. In this respect, it enunciates the general principles of national responsibility and it regulates a comprehensive liability regime for damage caused by a space object. Liability is assigned to the responsible party for the launch or return of a space object[650].

---

647 *Ibid.* Part 5 (76).

648 Nolon, *supra* note 48 at 685.

649 Australian Space Activities Bill 1998, Explanatory Memorandum.

650 The Australian law is silent with respect to the distribution of first party risks between the launch carrier and its customer. However, in light of the extraterritorial effect of foreign laws, particularly US and the characteristics of the launch market,

D. UNITED KINGDOM

1. NATIONAL APPLICATION OF INTERNATIONAL LAW

In the United Kingdom the Crown, as part of its Royal prerogative, has the power to negotiate and sign international treaties. By virtue of this same authority, the Crown also ratifies a treaty when ratification is necessary[651]. The actual exercise of the treaty making power falls with the Executive Branch of the government and is carried out by the Secretary for Foreign and Commonwealth Affairs. In this respect, Lord Atkin proclaimed in the Attorney-General for Canada v. Attorney-General for Ontario case that "within the British Empire there is a well-established rule that the making of a treaty is an executive act, while the performance of its obligations, if they entail alteration of the existing domestic law, requires legislative action[652]. However, even if the Parliament does not have any formal role in the treaty-making process, the British government has adopted a practice, known as the Ponsoby rule, under which the government submits any treaty requiring ratification to the Parliament for a period of twenty-one days before it is ratified[653].

Enrolled under the dualist doctrine, the United Kingdom demands the domestication of most international treaties for them to have legal effects at the national level[654]. Thus, the Parliament must necessarily transform the

---

the risk management approach in Australia will not differ radically from the ones followed in other jurisdictions. With respect to second party risks, the Act established a two-tiered risk allocation system, where the carrier must hire insurance or demonstrate financial responsibility up to the maximum probable loss and the Australian government assumes all the liability exceeding such maximum probable loss threshold. Third party risks are also allocated between the launch operator and the government on a horizontal basis, consisting of two layers. In the first one it is the launch operator that assumes all risks up to the amount of the insurance requirement. In the second layer the government assumes all risks.

[651] I. Sinclair, "National Treaty Law and Practice: United Kingdom" in M. Leigh & M. R. Blakeslee, eds., *National Treaty Law and Practice* (Washington, DC: ASIL, 1995) at 223.

[652] Attorney-General for Canada v. Attorney-General for Ontario [1937] A.C. 326 Lord Atkin.

[653] Lord Templeman, "Treaty Making and the British Parliament" in S. A. Riesenfeld & F. M. Abbott (eds.), *Parliamentary Participation in the Making and Operation of Treaties: A Comparative Study* (Dordrecht: Martinus Nijhoff Publishers, 1994) at 159.

[654] The European treaties are notably the most salient exception to this rule. Kindred, *supra* note 110 at 234.

international agreement by resorting to one of the existing domestication methods of an international treaty. In contrast, as decided in the Trendtex case, customary international law does not need transformation and is thus part of the law of the land[655].

In the event of inconsistencies between a statute and an international treaty, the British courts held that the statute should prevail over the treaty and that an international treaty included as a schedule of an implementing domestic statute is only a matter of interest and it may not detract from the natural meaning of the statute[656]. As emphasized by Brownlie, the doctrine of the prevalence of statutes over treaties is a constitutional principle and not a rule of construction. Nonetheless, a well-established rule of construction prescribes that where a national legislative act has been adopted to domesticate an international treaty there is a presumption that Parliament intended to fulfill its international obligations[657].

2. NATURE OF UNITED KINGDOM SPACE LAW

In 1986 the United Kingdom adopted an act on space activities. Its main purpose was to secure compliance with the international obligations of the United Kingdom with respect to the launching and operation of space objects and the carrying out of other activities in outer space by persons connected with Great Britain[658]. Since under UK law, the government has no general power to govern[659], the Parliament specifically conferred licensing and other powers to the Secretary of State, which were in turn delegated to the British National Space Center (BNSC), a partnership of the Department of Trade & Industry, the Foreign Office, the Office of Science and Technology and the Ministry of Defense. Before the establishment of the BNSC government responsibility for space endeavors was dispersed among several departments[660].

The Act consists of fifteen articles and it covers the licensing of activities, the license procedure, the register of space objects and a section of

---

[655] Trendtex Trading Corporation v. Central Bank of Nigeria [1977] I QB 529, CA.
[656] Ellerman Lines Ltd. V. Murray [1930- All E.R. 503 quoted by Lord Templeman, "Treaty Making and the British Parliament" in S. A. Riesenfeld & F. M. Abbott (eds.), *Parliamentary Participation in the Making and Operation of Treaties: A Comparative Study* (Dordrecht: Martinus Nijhoff Publishers, 1994) at 172.
[657] Brownlie, *supra* note 104 at 47.
[658] United Kingdom Act on Outer Space Activities, 1986 Chapter 38.
[659] F. Lyall, "UK Space Law", (1992) 35 *IISL* at 386.
[660] *Ibid.* at 387.

offences to the Act, among other aspects. It applies to the launch, procurement, and operation of a space object, whether carried on in the United Kingdom or elsewhere and to any activity in outer space. From a subjective standpoint, it applies to United Kingdom nationals[661], Scottish firms, and bodies incorporated under the law of any part of the United Kingdom.

By Order in Council Her Majesty may extend the application of the Act to bodies incorporated under the law of any of the Channel Islands, the Isle of Man or any dependent territory.[662] In 1996 it has been extended to Gibraltar and in 1998 to Cayman Islands to cover the launching operations of Sea Launch, a multinational firm incorporated under the laws of the Cayman Islands[663].

From a law reform perspective, the Secretary of State may make regulations prescribing anything required or authorized to be prescribed under this Act, and for carrying this Act into effect. Regulations under this Act must be made by statutory instrument which shall be subject to annulment in pursuance of a resolution of either House of Parliament. Nothing in the Act foresees the participation of interested parties in the adoption or annulment of regulations[664]. However, in 1989 the UK Parliamentary Space Committee (PSC) was formed to act as forum of discussion between members of Parliament and the space industry in order to promote a better understanding of space activities in the UK and the economic, technological and scientific benefits which they bring about[665]. The PSC is formed by members of Parliament of all political parties and representatives of the UK Industrial Space Committee[666] and the British Association of Remote Sensing

---

[661] United Kingdom national has been defined as an individual who is- (a) a British citizen, a British Dependent Territories citizen, a British National (Overseas), or a British Overseas citizen; (b) a person who under the British Nationality Act 1981 c. 61. 1981 is a British subject; or (c) a British protected person within the meaning of that Act.

[662] United Kingdom Act on Outer Space Activities, 1986 Chapter 38, article 2 (3).

[663] J. L. Reed , "The Commercial Space Launch Market and Bilateral Trade Agreements in Space Launch Services" (1997) *13 Am. U. Int'l L. Rev.* at 157.

[664] United Kingdom Act on Outer Space Activities, 1986 Chapter 38, article 11.

[665] M. Williamson, "The UK Parliamentary Space Committee. The emergence of a lobby", (1992) 8 *Sp Pol.* at 159.

[666] The United Kingdom Industrial Space Committee (UKISC) is the Trade Association of the British space industry. Founded in 1975, UKISC represents over three-quarters of the total turnover and employees of the industry. Its mission is to represent member companies within all parts of the UK space sector, and to promote the growth and competitiveness of the sector in the home and export markets.

Companies[667]. The PSC, which is the most active parliamentary space group in Europe, has been actively involved in the formulation of space policy and the identification of key priorities for UK space activities[668]. As arises from the preceding examination, the PSC ensures ample participation of the space industry in the elaboration of norms and it may even act as a control of the major decisions of the Secretary of State. Nonetheless, the general public does not have the possibility of participating in the committee.

The United Kingdom did not need to enact a large body of legislation, since its participation in space activities is channeled mainly through its involvement in ESA's programs[669]. The enactment of the Outer Space Act responded to internal constitutional requirements, since as arises from our above discussion international treaties do not have effect within national law[670] unless they are incorporated into UK law by Act of Parliament or by subordinate legislation[671].

3) RESPONSIBILITY AND LIABILITY

The Outer Space Act contains a set of provisions which aim at reallocating the liability imposed by the international Space Law instruments to the United Kingdom government. Unlike the US and the Australian regimes, the purpose sought by the United Kingdom is to reassign liability entirely to the licensee. It does not pursue a policy of promoting space activities or favoring the development of a certain space sector as is the case in other countries. Therefore, the United Kingdom does not assume any of the risks of its national space entities and thus the endeavors of the space industry do not entail any consequences to the UK taxpayer[672].

---

http://www.ukspace.com/trade/ukisc.htm accessed on July 6, 2001.

[667] Williamson, *supra* note 665 at 159.

[668] These include the following objectives: (i) to maintain UK's technology base at all costs, (ii) to maintain the UK lead in space communications, (iii) to encourage harmonized national and ESA programs for the exploitation of Earth-observation data, (iv) to encourage the European Union's market oriented approach, (v) to consider rejoining the European launcher program, and (vi) to reconsider, at a political level, manned space activities. Williamson, *supra* note 665 at 159.

[669] Lyall, *supra* note 659 at 385.

[670] This does not preclude UK responsibility at the international level. Brownlie, *supra* note 104.

[671] Lyall, *supra* note 659 at 385.

[672] *Ibid.* at 392.

3.1. First party risks

The Act does not contain any provisions dealing with the allocation of risks between the space launch carrier and its customer. Accordingly, the customer and the launch services provider are free to negotiate any risk sharing regime. However, as in the Australian case, the risk management approach may not deviate radically from the ones followed in other jurisdictions both on account of the negotiating power of the launch carrier and the extraterritorial effect of the US Commercial Space Launch Act.[673]

3.2. Second party risks

Since the United Kingdom is not engaged directly in the provision of launch related services second party risks, including international second party risks, are subsumed in the third party risk category.

3.3. Third party risks

The United Kingdom has reallocated all the third party risks to the licensee. For this purpose, the Act establishes a statutory indemnification whereby the licensee indemnifies Her Majesty's government against any claim brought against the government in respect of damage or loss arising out of activities carried on the licensee[674]. This indemnification applies even if the licensee is not the actual doer of the wrong or it is not the sole actor of the damage. For example, if the licensee is a satellite telecommunications operator and the launch vehicle of another State carrying its satellite causes damage to persons and property of a third State, this State may claim compensation from the United Kingdom in accordance with the Liability Convention[675] and the UK government may in turn seek the reimbursement of this compensation from the satellite telecommunications operator licensed under United Kingdom law even if the damage has been caused by the launch services provider. This is so because the United Kingdom would be a launching state and thus jointly and severally liable and there is no obligation under International Space Law for

---

[673] CSLA.

[674] United Kingdom Act on Outer Space Activities, 1986 Chapter 38, article 10 (1).

[675] The Convention prescribes that there is joint and several liability for damages caused when a space object is jointly launched by two or more states. In such case, the launching state which has paid compensation for damage is entitled to claim the proportional corresponding amounts to other participants in the joint launching. Thus, all launching states are equally liable for compensation unless they reach an agreement for a different division of liability.

the United Kingdom to seek reimbursement of the compensation from the other jointly and severally liable state[676].

In order to secure the availability of funds to face the obligations arising from the statutory indemnification the Secretary of State may require the licensee to insure itself against liability incurred in respect of damage or loss suffered by third parties as a result of the activities authorized by the license[677].

The indemnification does not apply to a person acting as employee or agent of another; or to damage or loss resulting from anything done on the instructions of the Secretary of State[678].

4. IMPLEMENTATION OF THE AUTHORIZATION AND SUPERVISION PRINCIPLE

The Outer Space Act establishes the general principle that all persons covered by the Act must obtain a license from the Secretary of State to carry out an activity to which the Act applies, i.e., the launch, procurement, and operation of a space object, whether carried on in the United Kingdom or elsewhere and any activity in outer space[679]. However, certain persons and activities have been exempted from the obligation to seek license. These include a person acting as employee or agent of another[680], and activities in respect of which the United Kingdom and another country have made arrangements to secure

---

676 Article V of the Convention expressly attributes joint liability for any damage caused by two or more States when they jointly launch a space object. It is possible under the Convention for participants in a joint launching to conclude agreements regarding the apportioning among themselves of the financial obligation in respect of which they are jointly liable. These agreements, however, may not impair the right of a state sustaining damage to seek the entire compensation due from any or all of the launching States. Also, in the absence of said agreement, the State which has paid compensation for damage is entitled, but not obliged, to present a claim for indemnification to other participants in the joint launching.

677 United Kingdom Act on Outer Space Activities, 1986 Chapter 38, article 5 (2).

678 Francis Lyall disagrees with this provision as he envisages a scenario where a satellite operator is required by the government to alter orbit in order to reduce space debris. Lyalls wonders why the UK might be liable for damage but the operator and even the insurer will be free from the duty to compensate damages. Lyall, *supra* note 659 at 387. United Kingdom Act on Outer Space Activities, 1986 Chapter 38, article 10 (2).

679 United Kingdom Act on Outer Space Activities, 1986 Chapter 38, article 3 (1).

680 *Ibid.* Chapter 38, article 3 (2) (a).

compliance with the international obligations of the United Kingdom[681]. The latter requires certification of the Order in Council. The Secretary of State may also exempt other persons or activities if satisfied that the requirement is not necessary to secure compliance with the international obligations of the United Kingdom[682]. Additionally, since the requirements for the obtainment of a license revolve around the concept of nationality, it has been noted that intergovernmental and international organizations, such as IMSO, which has been established as a United Kingdom firm[683], do not need to obtain a UK license to carry out activities in space[684].

The Act vested the Secretary of State with ample discretionary powers to grant a license[685] and to impose conditions to the license[686]. However, in light of our findings with respect to the international constraints and faculties of governments in the adoption of domestic space law, these discretionary powers must respect the doctrines and principles of international law. Thus, the Secretary of State must exercise this power in good faith, reasonably, and in conformity with the spirit of the Outer Space Treaty[687].

The Act contains general standards for the award of the license[688]. In this respect, the Secretary of State must ensure that the activities will not jeopardize public health or the safety of persons or property, that they will be consistent with the international obligations of the United Kingdom, and that they will not impair the national security of the United Kingdom[689].

The Act does not establish the procedure for the award of the license. It confers the Secretary of State the power to make regulations to adopt the procedure to be followed and to prescribe the form and contents of license applications[690].

---

681 *Ibid.* Chapter 38, article 3 (2) (b).

682 In these cases, an order must be made by statutory instrument, subject to annulment in pursuance of a resolution of either House of Parliament. United Kingdom Act on Outer Space Activities, 1986 Chapter 38, article 3 (3).

683 D. Sagar, "Inmarsat since Privatization", Project 2001, Working Group on Telecommunication, at 163.

684 F. Lyall, *supra* note 659 at 387.

685 United Kingdom Act on Outer Space Activities, 1986 Chapter 38, article 4 (1).

686 *Ibid.* Chapter 38, article 5.

687 See *supra* Chapter I.

688 United Kingdom Act on Outer Space Activities, 1986 Chapter 38, article 4 (2).

689 *Ibid.*

690 *Ibid.* Chapter 38, article 4 (3).

A license may contain certain conditions[691], such as a requirement to conduct operations so as to prevent the contamination of outer space or adverse changes in the environment of the earth, to avoid interference with the activities of others in the peaceful exploration and use of outer space, to avoid any breach of the United Kingdom's international obligations, and to preserve the national security of the United Kingdom[692]. The Secretary of State may also impose the licensee certain conditions with respect to the disposal of the payload in outer space on the termination of operations under the license[693].

4.1. Satellite telecommunications licenses

Satellite telecommunications activities in the United Kingdom are governed by a separate set of rules[694]. These may need a license under one of three separate regimes: (i) the Wireless Telegraphy Act[695], which regulates spectrum use, (ii) the Telecommunication legislation, which regulates the operation of systems and the provision of telecommunications services[696] and, (iii) broadcasting legislation, which in very general terms regulates content[697].

The 1949 Wireless Telegraphy Act conferred powers to the Secretary of State to award licenses for the use of the civil radio spectrum[698]. Licenses are grouped under three major classes: (i) the pre-packed license product; (ii) customized license product; and (iii) spectrum license products. The pre-packed license product includes standardized conditions of use, with pre-determined frequencies[699]. Customized license product applies where

---

[691] *Ibid.* Chapter 38, article 5.
[692] *Ibid.* Chapter 38, article 5 (2).
[693] *Ibid.* Chapter 38, article 5 (2).
[694] D. Gillies & R. Marshall, *Telecommunications Law* (London: Butterworths, 1997) at 613. This subsection derives mainly from D. Gillies & R. Marshall, *Telecommunications Law* (London: Butterworths, 1997); S. White, Stewart, S. Bate & T. Johnson, *Satellite Communications in Europe: Law and Regulation* (London: FT Law & Tax, 1996) and the proceedings of *QDOS Telecommunications Law*, London, 9-10 February, 1998.
[695] Wireless Telegraphy Act, 1998, (Eng.).
[696] Telecommunications Act, 1984, (Eng.).
[697] Gillies & Marshall, *supra* note 694 at 613.
[698] Licenses are issued by the Radiocommunications Agency, an executive agency of the Department of Trade and Industry. Its main responsibility is to effectively manage the radio spectrum. *Ibid.* at 614.
[699] *Ibid.* at 613. According to Foley: "often this spectrum is agreed at the International Telecommunication Union for similar use throughout the world. There is a simple scale fee for such use since normally many users are sharing the same pre-packaged spectrum. Examples include ship or aircraft radio licenses or short range business

applicants have a specific requirement which they have determined is unique to their business[700]. In this case a detailed technical processing is required to tailor the frequency concerned so that it will not interfere with other users[701]. These licenses involve site or frequency clearing. Finally, spectrum license products are licenses where a block of spectrum or set of predefined channels are agreed before the license is issued. This form of license granted to a self-management organization is unique to the United Kingdom[702].

The Telecommunications Act of 1984[703] prescribes that those individuals or entities who run[704] telecommunications systems need a license[705] and makes non compliance with this norm a criminal offense. The act provides for two main categories of licenses to be granted: (i) class licenses or general authorizations, and (ii) individual licenses. Class licenses do not actually require any authorization and are automatically considered granted if the projected telecommunications system falls within the provisions of a class license. There are five main class licenses (i) the self-provision license, (ii) the telecommunications services license, (iii) the private mobile radio class license, (iv) the satellite services class licenses and, (v) the cordless class license[706]. The satellite services license allows the running of satellite transmit and/or receive terminals of any kind for the provision of a wide range of services, provided that the satellite transmitting and receiving terminals are not connected, directly or indirectly, to the public switched network[707]. However if an earth station at the far end is authorized for connection to the public switched network, it is possible to run a remote earth

---

classes where there is no customized frequency assignment". Foley, Paul, "Trends in Spectrum Management - with Specific Emphasis on Spectrum Pricing Considerations", *QDOS Telecommunications Law*, London, 9-10 February, 1998 at 3.

[700] Gillies & Marshall, *supra* note 694 at 613.

[701] *Ibid.* at 613.

[702] For example, in the case of some major radio based networks the Radiocummunications Agency does not assign specific frequencies for a particular station, but instead provides a block of spectrum to the user. The license holder then plans the location of stations within the network and the use of channels by those stations within the block of assigned spectrum. Foley, *supra* note 699 at 3.

[703] United Kingdom Telecommunications Act of 1984. Several amendments were introduced to the 1984 Act through the Licensing Regulations issued in October 1997 which implemented the Licensing Directive. SI 1997/2930.

[704] The term "run" has not been defined by the act or the regulations. Gillies & Marshall, *supra* note 694 at 613.

[705] Service providers who do not run any system do not require a license.

[706] United Kingdom Telecommunications Act of 1984.

[707] C. Raikes, "Licensing Trends in Telecoms - The DTI Policies and Practices", *QDOS Telecommunications Law*, London, 9-10 February, 1998 at 1.

station under the satellite services class license. Messages from a mobile or transportable earth station which are intended to be received by an overseas downlink connected to the public switched network are also permitted under the license. Individual licenses are granted to individuals or entities intending to run a system for the provision of services falling outside the various class licenses.[708]

The Broadcasting Act of 1990, as amended in 1996[709], established the obligation for television and radio services provided from the United Kingdom to be licensed and created a general legal framework aimed at regulating television program and additional services[710]. With respect to satellite television services, the act distinguishes between domestic and non domestic satellite services[711]. The former are defined as a television broadcasting service where the television programs included in the service are transmitted on an allocated frequency from a place in the United Kingdom for general reception in the United Kingdom. The latter in turn are defined as the transmission of television programs by satellite other than an allocated frequency from a place in the United Kingdom for general reception in either or both the United Kingdom or any prescribed country.[712] Both domestic and non domestic satellite services require a license provided by the Independent Television Commission. Non-domestic satellite service licenses are subject to a fee and to the Commission's verification that the licensee can comply with the license conditions. Domestic satellite services are subject to a more discretionary licensing procedure.[713]

---

[708]*Ibid.* at 1.

[709] Broadcasting Act, 1990, amended by Broadcasting Act, 1996(Eng.).

[710]According to the act, "the essence of a broadcast is that it is simultaneously made available to more than one person by means of a telecommunications system. The act also created three bodies: the Independent Television Commission, the Radio Authority and the Broadcasting Standards Commission. The first two have virtually identical duties with regard to the regulation of the content of broadcast services, the first in relation to television and the second to radio. Both are empowered to issue licenses in respect of broadcast services. The Independent Television Commission is charged with the function of regulating the provision of television program services, local delivery services and added digital services". Gillies & R. Marshall, *supra* note 694 at 613.

[711]White, Bate & Johnson, *supra* note 694 at 227.

[712] *Ibid.* at 227.

[713]*Ibid.* at 228. Raikes, *supra* note 707 at 1.

4.2. Continuing supervision

There is a comprehensive system to enforce the continuing supervision obligations. First, the Secretary of State may inspect the licensee's facilities, and may even inspect and test the licensee's equipment[714]. Second, the Secretary of State may require the licensee to provide information regarding the launch and the nature, conduct and results of the licensee's activities. To carry out these tasks, the Secretary of State is entitled to inspect and take copies of documents relating to this information[715].

Additionally, the Act granted the Secretary of State power to give directions to any person acting in contravention of the licensing requirements or conditions to rectify that conduct so as to secure compliance with the international obligations of the United Kingdom or so that the licensee will comply with the conditions of the license[716]. Furthermore, the license is also subject to variation, suspension and termination when a condition or any regulation has not been complied with, or if it is required in the interests of public health or national security, or to comply with any international obligation of the United Kingdom[717].

The Act also implemented a system of criminal offences for those that carry out an activity in contravention of the licensing requirements, the regulations adopted under the Act and the directions of the Secretary of State[718].

5. IMPLEMENTATION OF THE REGISTRATION OBLIGATIONS

The Secretary of State must maintain a register of space objects, containing all the particulars deemed appropriate to comply with the international obligations of the United Kingdom. For this purpose, the Secretary of State may require the licensee to provide information as to the date and territory or location of the launch, the basic orbital parameters, including nodal period, inclination, apogee and perigee, and any information deemed necessary

[714] United Kingdom Act on Outer Space Activities, 1986 Chapter 38, article 5 (2).
[715] *Ibid.* Chapter 38, article 5 (2).
[716] *Ibid.* Chapter 38, article 8.
[717] The suspension, revocation or expiry of a license does not affect the obligations of the licensee under the conditions of the license. United Kingdom Act on Outer Space Activities, 1986 Chapter 38, article 5 (2).
[718] United Kingdom Act on Outer Space Activities, 1986 Chapter 38, article 12.

concerning the nature, conduct, location and results of the licensee's activities[719].

A unique feature of the United Kingdom legislation is the fact that it maintains a supplementary registry of space objects. This has been created to record those space objects whose launch has been procured by a UK satellite supplier, but which appear on the registry of another State party to the Registration Convention. The supplementary registry also records those satellites whose title and control has been transferred to a UK satellite operator after launch and the Secretary of State has licensed that company to operate the satellite. This supplementary system is considered to give transparency on all the licenses issued under the Outer Space Act 1986[720].

6. CONCLUDING REMARKS

The United Kingdom has enacted a single legislative instrument to govern space activities. The main purpose is to comply with the obligations assumed by the United Kingdom at the international level. The Act confers ample discretionary powers to the Secretary of State to issue licenses, impose conditions and supervise activities. From a law reform perspective, the Secretary of State may make regulations which are subject to annulment in pursuance of a resolution of either House of Parliament. Although the Act does not foresee the direct participation of interested parties, a Parliamentary Space Committee was formed to act as forum of discussion between members of Parliament and the space industry. This committee ensures ample participation of the space industry in the formulation of policy and it is the most active space parliamentary committee in Europe.

The Outer Space Act adopted a statutory indemnification to reallocate to the licensee third party liability imposed to the UK government under international Space Law treaties and conventions. Consequently, the United Kingdom does not assume any of the risks of its national space entities.

The Act has also established a comprehensive licensing system for space activities. The licensing of satellite telecommunications services is effectuated under a sophisticated set of specific regulations.

The United Kingdom has adopted a unique dual register of space objects composed of a main and a supplementary registry. The latter one

---

[719] *Ibid.* Chapter 38, article 7.

[720] http://www.bnsc.gov.uk/index.cfm?nid=9935 accessed on June 20, 2001.

records all those space objects whose launch has been procured by a UK satellite supplier, but which appear on the registry of another State.

E. UKRAINE

1. NATIONAL APPLICATION OF INTERNATIONAL LAW

According to the recently adopted Constitution, the President represents the state in international relations, administers foreign policy[721], conducts negotiations and concludes international treaties. The Verkhovna Rada, Ukraine's parliament, is entrusted with approving international treaties[722]. Once the Verkhovna Rada grants its consent, international treaties become part of the national legislation of Ukraine, binding the courts, the Government and private persons. The Parliament also has the faculty of denouncing international treaties[723].

With respect to the hierarchy of international law vis-à-vis domestic law, the 1996 Constitution only provides that "international treaties currently in force, as ratified by the Supreme Rada of Ukraine, form part of Ukraine's national legislation." Therefore, at the constitutional level Ukraine did not proclaim that international treaties take priority over contrary domestic legislation[724]. The supremacy of certain international treaties over contrary Ukrainian legislation is established only in the 1993 Law on International Treaties of Ukraine[725]. Art. 17 of this law provides that "if the international treaty of Ukraine, concluded in the form of a law, establishes other rules than those provided in the legislation of Ukraine, then those applied shall be the rules of the international treaty[726]."

---

[721] The foreign policy of Ukraine is aimed at ensuring its national interests and security by maintaining peaceful and mutually beneficial co-operation with members of the international community, according to generally acknowledged principles and norms of international law. Ukraine Constitution article 18.

[722] Ukraine Constitution article 85.32.

[723] *Ibid.* article 85.32.

[724] In contrast, the "generally recognized principles and norms of international law" are mentioned only in the clause dealing with foreign policy. Art. 18 of the Ukrainian Constitution provides that "the foreign political activity of Ukraine is aimed at ensuring its national interests and security by maintaining peaceful and mutually beneficial co-operation with members of the international community, according to generally acknowledged principles and norms of international law".

[725] G. M. Danilenko, "Implementation of International Law in Russia and Other CIS States," 1998 [unpublished] at 36.

[726] *Ibid.* at 37.

The Constitutional Court[727] is authorized to provide opinions on the conformity of international treaties in force and international treaties submitted to the Parliament with the Constitution of Ukraine[728]. These faculties, which include the possibility of declaring a treaty to be unconstitutional, underscore the supremacy of national law over international treaties[729].

2. NATURE OF UKRAINIAN NATIONAL SPACE LAW

After the disintegration of the former Soviet Union, Ukraine inherited a significant stake of the Soviet Union's space industry and it found it necessary to adopt specific legislation to govern the development of space activities[730]. Thus, in 1996 it enacted the law on Space Activities, which was complemented by other legislative acts established in conformity with this law[731]. The law sets the fundamental principles of the organization of space activities and outlines the basic principles of State regulation and management of space activities[732]. It creates the Ukrainian National Space Agency, the central executive authority responsible for implementing State policy in relation to space activities, and it determines the competences of the agency[733].

---

[727] On the appeal of the President of Ukraine or the Cabinet of Ministers of Ukraine. Ukraine Constitution, article 151.

[728] Ukraine Constitution, article 151.

[729] H. Hannum, "The Status and Future of the Customary International Law of Human Rights: The Status of the Universal Declaration of Human Rights in National and International Law" (1995/1996) 25 *Ga. J. Int'l & Comp. L.* at 287.

[730] R. Krawec, "Ukrainian Space Policy Contributed to Natural Economic Development", (1995) 11 *Sp. Pol.* at 105.

[731] Ordinance of The Supreme Soviet Of Ukraine, on Space Activity, Law of Ukraine of 15 November 1996 (VVRU, 1997, p. 2) [hereinafter the "Law Ukraine on Space Activity"].

[732] The Law Ukraine on Space Activity, article 5.

[733] The Ukrainian National Space Agency shall, within its competence: formulate the conceptual basis of State space policy; provide for the organization of space activity in Ukraine and under the jurisdiction of Ukraine outside its borders; prepare, the National Space Program and ensure its implementation; direct the management and coordination of the work of enterprises, institutions and organizations in the space and related sectors; act as the general State customer, manufacture and test of space technology; arrange for the operation, maintenance and improvement of space facilities; arrange for licensing of space activity and the development and operation of the Ukrainian Space Technology Certification System; register space technology; and promote international cooperation; and participate in the preparation of international treaties. The Law Ukraine on Space Activity, article 6.

Additionally, it lays down the basis for Ukraine's participation in international cooperation endeavors[734] and it creates a fund for the financing of national space activities[735]. The law, which has been drafted in very broad and ambiguous terms, refers many of the most significant legal aspects of space activities, such as the scope of the licensing, certification and registration of space activities, the supervision of the safety of operations and the standards for the environmental protection, to the regulations, which have to be adopted by different State authorities of Ukraine within their competence.[736] Despite the importance of the issues referred to the regulations, the law does not foresee any mechanism for the participation of interested parties in the elaboration of these or other regulations.

3. RESPONSIBILITY AND LIABILITY

There is no mechanism for the redistribution of risks imposed to the State by the international Space Law instruments. Furthermore, the Law does not contain any provisions on responsibility and it only states that "liability for damage sustained in the course of a space activity, as well as procedures for determining the extent of such damage" will be established by the adoption of future legislation[737]. Similarly, all aspects of insurance to be procured in connection with the pursuit of space activities will also be established by legislation[738]. This may be attributed to the fact that Ukraine's participation in outer space activities is rather limited in comparison with major spacefarers, such as the United States, Russia, and France.

4. IMPLEMENTATION OF THE AUTHORIZATION AND SUPERVISION PRINCIPLE

Ukraine has implemented a dual licensing and registration system to comply with the authorization principle. In this respect, it prescribes that any space facility engaging or intending to engage in space activity in Ukraine or under the jurisdiction of Ukraine outside its borders must obtain a license from the Ukrainian National Space Agency[739]. Space facility is defined as any object designed, manufactured and operated for the purpose of exploring and using outer space[740]. Unlike most other jurisdictions, Ukraine requires space

---

[734] *Ibid.* article 17.
[735] *Ibid.* article 11.
[736] *Ibid.*. article 8.
[737] *Ibid.* article 25.
[738] *Ibid.* article 24.
[739] *Ibid.* article 10.
[740] *Ibid.* article 1.

facilities, i.e., space technology and objects, to be authorized rather than the activities of non governmental entities. This is not in itself incompatible with the international regime[741]. However, this may create some obstacles for the smooth development of the private space industry. The procedures for the licensing of space activities in Ukraine are established by the Cabinet of Ministers of Ukraine and are not contained in the law[742].

Space facilities in Ukraine are also subject to certification attesting to its compliance with operating requirements established by the regulations. This process leads to the issuance of a compliance certificate[743]. The procedure is determined by the Ukrainian Space Technology Certification System[744].

Apart from the licensing and registration obligations, the law prescribes that all those engaged in space activities must comply with safety requirements with regard to the life and health of the public, the property of citizens, enterprises, institutions and organizations and the protection of the environment and must take all necessary measures to prevent environmental damage[745]. However, the law does not specify any safety parameters, which are delegated to future regulations, or any definition of environmental damage.

Another distinctive feature of the Act is the fact that it conditions the conduct of space activities to certain principles[746]. It does not determine the consequences of not following these principles, but it may be construed that a space activity which does not fall within these principles would not be licensed. These principles[747], which have been formulated in broad terms, are State regulation, progressive development of State space policy, practical exploitation of the scientific and technical potential of Ukraine in the interest of the national economy, scientific advancement and State security and for commercial purposes and promotion of international cooperation[748].

---

[741] See *supra* Chapter I.
[742] *Ibid.* article 10.
[743] *Ibid.*
[744] *Ibid.*.
[745] *Ibid.* article 21.
[746] *Ibid.* article 4.
[747] *Ibid.*
[748] *Ibid.*

4.1. Continuing supervision

The law does not specify the way the government implements the continuing supervision obligations. It simply states that supervision of compliance with safety requirements and other space regulations is the responsibility of the Ukrainian National Space Agency, the Ministry of Defense of Ukraine and other executive authorities within their competence.

5. IMPLEMENTATION OF THE REGISTRATION OBLIGATIONS

Unlike all other jurisdictions[749], the law provides for the registration of space facilities instead of space objects. The concept of space facilities is broader than that of space objects, since it encompasses not only space objects which are designed, manufactured and operated in outer space, but also objects that are operated exclusively on the Earth surface for the purpose of exploring and using outer space, such as ground segments and ground infrastructure. Therefore, all space facilities must be recorded in the State Register of Space Facilities. [750]

The law exempts from registration those space facilities which have been designed with corporate entities of other countries or international organizations, in which case the registration is decided in accordance with the international agreements concluded[751].

The law does not contain any information which the interested party must provide to the governmental authorities for the registration of space facilities.

6. CONCLUDING REMARKS

The law on Space Activities adopted in 1996 is the sole legislative instrument in Ukraine which governs substantive domestic issues in Ukraine. This law, which is drafted in very broad and imprecise terms, refers many of the most significant legal aspects of space activities to the regulations, which have to be adopted by different State authorities. Despite the importance of the issues referred to the regulations, the law does not foresee any mechanism for the participation of interested parties in the elaboration of these norms.

---

[749] See *supra* Chapter II.
[750] The Law Ukraine on Space Activity, article 13.
[751] *Ibid.* article 13.

Due to Ukraine's relatively limited participation in outer space activities, there is no mechanism for the redistribution of risks arising from the international Space Law instruments. All liability and insurance issues have been referred to future legislation.

Compliance with the authorization principle has been implemented through a double licensing and registration system of space facilities. The law also obliges the abidance by safety and environment requirements, but it does not specify any actual parameters, which are delegated to future regulations. The Act conditions the legitimacy of space activities to certain principles and goals, all of which have been formulated in very broad terms. The continuing supervision obligation has been entrusted mainly to the Ukrainian National Space Agency and the Ministry of Defense of Ukraine. However, the law does not specify how these state organisms will carry out their supervision obligations.

Under Ukrainian law, space facilities, which include not only space objects but also ground infrastructure, must be recorded in the State Register of Space Facilities. There is no indication of the information which must be provided in connection with the registration of space facilities.

F. SWEDEN

1. NATIONAL APPLICATION OF INTERNATIONAL LAW

According to the Swedish Constitution, the Government, which exercises executive functions, concludes international agreements with other States and international organizations[752]. However, an international agreement that presupposes the amendment or abrogation of a law or the enactment of a new law needs Parliament approval[753]. Treaties of major importance also require parliamentary approval. The Government must keep the Foreign Affairs Advisory Council, which is composed of the Speaker and other nine members of Parliament, continuously informed of any foreign affairs matter of importance. In all foreign policy matters of major importance, the Government is expected to seek the Council's viewpoint before making its decision[754].

---

[752] Swedish Constitution, chapter 10, article 1.
[753] *Ibid.* article 2.
[754] *Ibid.* article 6.

Sweden follows the dualist approach to international treaties. Thus, implementation of an international treaty is required for making its provisions applicable by Swedish courts and public authorities. International obligations under a ratified treaty do not automatically prevail over the Constitution[755].

2. NATURE OF SWEDISH NATIONAL SPACE LAW

Swedish domestic space law consists of a single non-comprehensive statute, complemented by a regulatory decree. The act, denominated Act on Space Activities, was adopted in 1982 to comply with the obligations assumed by Sweden at the international level[756]. It is concise and does not even purport to formulate national space policy or to create mechanisms for the future elaboration of national norms. It does not provide any mechanism for participation of those affected by the law to be involved in the elaboration of regulations.

The Act applies to activities in outer space, referred to as space activities, and to the launching of objects and all measures to operate or affect objects launched into outer space[757]. However, the mere reception of satellite signals is not designated as space activities[758].

3. RESPONSIBILITY AND LIABILITY

The Act reallocates all liability which may be imposed to the Swedish State on account of its international undertakings to the persons who have carried on the space activity[759]. There is no clear indication of the concept of "person carrying out a space activity", but it may be construed in light of other provisions of the act that this concept refers to the licensee.

Additionally, the law prescribes the possibility of creating exceptions for the assumption of all liability by the licensee under special reasons[760]. Nothing in the act or the regulations explains the extent of those special

---

[755] Council of Europe, The Implications for Council of Europe Member States of the Ratification of the Rome Statute of the International Criminal Court, Progress Report, Sweden, consult/icc (2001) 37.

[756] Swedish Act on Space Activities (1982:963).

[757] *Ibid.* section 1.

[758] The launch of sounding rockets is not considered space activities either. *Ibid.* section 1.

[759] Swedish Decree on Space Activities (1982:1069) section 6.

[760] *Ibid.* section 6.

reasons. Neither do they provide a standard for the determination of those reasons.

4. IMPLEMENTATION OF THE AUTHORIZATION AND SUPERVISION PRINCIPLE

The Act prescribes that space activities may not be carried out from Swedish territory without a license. Furthermore, Swedish natural or juridical persons also need a Swedish license to carry out space activities anywhere else[761]. However, the act does not define the legal grounds for the award of a license.

A license to carry on space activities is granted by the National Board for Space Activities. The decree establishes the basic requirements for the application of the license but it does not implement a clear procedure[762]. The Board is directed to consult the telecommunications administration and other national ministries or authorities affected by the application[763]. Nonetheless, no clear indication of the authorities or the extent of the consultation has been given.

The domestic instruments permit the governmental authorities to restrict license in the way deemed appropriate[764]. Neither the act nor its regulations establish clear parameters for the restrictions of the rights of the licensee. Additionally, the Act prescribes that any person who carries out space activities without a license or disregards its conditions, is subject to criminal sanctions[765].

4.1. Continuing supervision

The Swedish norms implemented a system of inspection of the space activities of the licensee to comply with the continuing supervision obligations of the Swedish State[766]. Additionally, the National Space Board may impose conditions to the license to facilitate the control of the space activities. Another measure adopted to fulfill the supervision requirements is the possibility of withdrawing a license if its conditions have been disregarded[767].

---

761 Swedish Act on Space Activities (1982:963) section 2.
762 Swedish Decree on Space Activities (1982:1069) section 1.
763 *Ibid.* section 1.
764 Swedish Act on Space Activities (1982:963) section 3.
765 Swedish Act on Space Activities (1982:963) section 5.
766 *Ibid.* section 3.
767 Pending a final decision on its withdrawal, a license may be withdrawn

5. IMPLEMENTATION OF THE REGISTRATION OBLIGATIONS

The National Board for Space Activities established a register of objects launched to outer space to record objects in accordance with Article 1 of the Registration Convention and for those joint launches where the States involved designated Sweden for the registration of space objects[768].

Like most of other jurisdictions, the information to be recorded follows the standards contained in article IV of the Registration Convention. Thus, the register records the designation or registration number for the space object, which is assigned by the National Space Board through the register itself, the date and territory or location of launching, the basic orbital parameters and a general function of the space object[769].

The Board acting through the Ministry for Foreign Affairs must provide the Secretary General of the United Nations with information recorded in the register as provided for in article IV of the Convention[770].

6. CONCLUDING REMARKS

The Swedish Act is restricted to implementing certain measures in order to comply with the international obligations assumed by Sweden at the international level. It is succinct and does not even elaborate or provide the basis for the elaboration of future space policy. From a participatory theory perspective, it does not contemplate the participation of the industry or the general public in the elaboration of regulations.

The norms dealing with responsibility and liability are insufficient and imprecise and tend to reallocate the risks to the license holder, unless the authorities consider it appropriate to adopt other solutions if there are special reasons to do so. No clarification whatsoever of the standard of special reasons have been given in the Act or its regulatory decree.

The Act implemented a licensing system for space activities carried out from Swedish territory and for Swedish entities and individuals conducting space activities in foreign jurisdictions. However, the act does not define the legal grounds for the award of a license and does not establish any

---

temporarily.

[768] Swedish Act on Space Activities (1982:963).

[769] Swedish Decree on Space Activities (1982:1069) section 4.

[770] *Ibid.* section 4.

procedure, which gives ample discretionary powers to the National Board for Space Activities. The same conclusions apply to the adopted continuing supervision system.

The implementation of the registration obligations follows the standards and parameters established in the international Space Law instruments.

G. SOUTH AFRICA

1. NATIONAL APPLICATION OF INTERNATIONAL LAW

The Constitution of South Africa places the treaty-making power in the hands of the National Executive –headed by the President- which is entitled to negotiate and sign international agreements[771]. However, the Constitution confers a control function to the Parliament by providing that an international agreement[772] is binding on South Africa only after it has been approved by resolution in each house of Parliament, i.e., the National Assembly and the National Council of Provinces[773].

Like most common law jurisdictions, South Africa follows the dualist or "transformationst" approach with respect to international law. Therefore, an international agreement becomes law in South Africa only "when it is enacted into law by national legislation"[774]. Nonetheless, the Constitution mandates every court, when interpreting any legislation, to prefer any reasonable interpretation that is consistent with international law over any alternative construction[775]. Also, a self-executing provision of an agreement approved by Parliament is law unless inconsistent with the Constitution or an Act of Parliament[776]. Similarly, customary international law is law in South Africa unless it is inconsistent with the Constitution or an Act of Parliament[777].

---

[771] Constitution of the Republic of South Africa, section 231 (1).

[772] The following treaties are exempted from parliamentary approval: agreements which are technical, administrative or executive in nature. Constitution of the Republic of South Africa, section 231 (3).

[773] *Ibid.* section 231 (2).

[774] *Ibid.* section 231 (4).

[775] *Ibid.* section 233. Specifically, when interpreting the Bill of Rights the courts are required to "consider international law," and "may consider foreign law."

[776] Constitution of the Republic of South Africa, section 231 (4).

[777] *Ibid.* section 232.

International law has also influenced the debate over the adoption of both the Interim and the Final Constitutions and it plays a significant role in the decisions of the courts.[778]

2. NATURE OF SOUTH AFRICAN NATIONAL SPACE LAW

In 1993 South Africa enacted the Space Affairs Act[779], whose one of the main purposes –among several others- is the establishment of the Council for Space Affairs, a state organism under the authority of the Department of Trade and Industry, which is responsible for the implementation of South African space policy.[780] The Act prescribes that the Minister of Trade and Industry, in consultation with the Council for Space Affairs, must determine the general space policy with a view to meeting all the international commitments and responsibilities of South Africa in the space field so as to be recognized as a responsible and trustworthy user of outer space[781]. The other major policy purpose is the control and restriction of the development, transfer and acquisition of dual-purpose technologies, in light of the international arrangements adopted by South Africa[782].

The Act foresees the possibility of the Minister's delegating to the chairperson of the Council or any officer of the Department of Trade and Industry the powers conferred under the Act, with the exception of the obligation to hear appeals and make regulations[783]. The Minister may adopt regulations regarding the licensing procedures, measures to protect the national interests, safety measures and minimum safety standards concerning any space activity, among other issues contained in the Act[784].

The Act does not contemplate the possibility of public consultations for the adoption of new regulations. However, it devised a mechanism which affords many opportunities for the space industry and the general public to be actively involved in the different aspects of the implementation of space

---

[778] J. Sarkin, "The Effect of Constitutional Borrowings on the Drafting of South Africa's Bill of Rights and Interpretation of Human Rights Provisions" (1998) 1 *U. Pa. J. Const. L.* at 184.
[779] South African Space Affairs Amendment Act, No. 84 of 2 July 1993.
[780] The Act was amended to incorporate minor changes in 1995. South African Space Affairs Amendment Act, No. 64 of 6 October 1995.
[781] South African Space Affairs Amendment Act, No. 84 of 2 July 1993, article 2.
[782] *Ibid.* article 2.
[783] *Ibid.* article 5 (3).
[784] *Ibid.* article 5 (3).

policy. In this respect, the Council is instructed to hear representations regarding space affairs by any person, to encourage persons and authorities involved in the space industry to register with the Council, to designate knowledgeable persons from the space industry as members to the committees of the Council, with the view to assisting the Council in the performance of its functions, and to encourage the responsible participation by any person in space affairs[785]. Furthermore, the Council is integrated with two persons from the space industry.[786]

3. RESPONSIBILITY AND LIABILITY

The Act does not establish any substantive decision regarding responsibility and liability. Therefore, absent a specific regulation or the imposition of an express condition, it does not reallocate the liability assumed by the South African government under the international treaties and conventions.

However, the Act expressly authorizes the Council and the Minister of Trade and Industry to issue conditions to the license to determine a liability regime[787]. In particular, these liability conditions may determine the risks to be assumed by the licensee and may even limit or exclude the liability of the licensee for damage caused in connection with the activities covered by the license, whether caused by the licensee's fault or not. The conditions may also impose the licensee to give security to meet the obligations that may be incurred. The law is silent as to what type of security it may be required. But a wide interpretation of its relevant provisions leads to the conclusion that all kind of risk administration tools and strategies would be acceptable, such as such as insurance, bonds, self-insurance, and risk retention, among others[788].

4. IMPLEMENTATION OF THE AUTHORIZATION AND SUPERVISION PRINCIPLE

South Africa adopted a licensing system to comply with the authorization and supervision obligations assumed at the international level. In this sense, it prescribes that no person may perform any of the activities contemplated in the act except with a license duly issued by the Council for Space Affairs[789].

---

[785] *Ibid.* article 5 (3).
[786] *Ibid.* article 6 (3).
[787] *Ibid.* article 14 (2).
[788] C. A. Williams, Jr., M. L. Smith, & P. C. Young, *Risk Management & Insurance*, 7th ed. (New York: McGraw-Hill Inc., 1995) at 9.
[789] South African Space Affairs Amendment Act, No. 84 of 2 July 1993, article 11 (1).

The activities covered by the Act include any launching from the South African territory, or from the territory of another state by a legal entity incorporated in South Africa, the operation of a launch facility, and the participation by legal entities incorporated in South Africa in space activities entailing international obligations to the South African State or which may affect national interests[790]. Additionally, any other space or space-related activities prescribed by the Minister will require a South African license[791].

The Council is entitled to impose conditions to a license, taking into account the minimum safety standards, the national interests and the international obligations and responsibilities. The Act also delineates the general outlines for the license procedure. Nonetheless, it does not contain specific standards for the award of the license, which gives ample discretion to the Council. It does, however, contemplate the possibility to appeal any decision of the Council to the Minister[792] and to resort to a court of law[793], for which cases the Act delineates a clear procedure.

4.1. Continuing supervision

In order to comply with the continuing supervision obligations, the Council may designate inspectors, who are entitled to enter any facility of the licensee or any facility where the Council believes that an activity covered by the Act is carried out without a license[794]. The inspectors may also be present at any activity of the licensee to determine whether it poses an unacceptable safety risk. The inspectors may further audit any document concerning the license[795].

Additionally, the Council is entitled to amend, suspend and revoke the license once it has been issued in the event that the licensee violates any condition or regulation or whenever the continuation of a licensed activity may impose an unacceptable safety risk[796]. In this case, the Council may give directions to the licensee in order to prevent the occurrence of damage[797]. The non compliance with a condition of the license and other violations of the law and the regulation, such as the performance of a space activity without a

---

790 *Ibid.* article 11 (1).
791 *Ibid.* article 11 (1).
792 *Ibid.* article 16.
793 *Ibid.* article 17.
794 *Ibid.* article 10 (1).
795 *Ibid.* article 10 (4).
796 *Ibid.* article 13.
797 *Ibid.* article 14 (5).

license, constitute a criminal offense punished with fines and even imprisonment[798].

5. IMPLEMENTATION OF THE REGISTRATION OBLIGATIONS

The law does not contemplate any registration of space objects or the creation of a national space registry. This is because so far South Africa has not launched any space object and it has not had the obligation to provide the United Nations with information regarding any object launched into outer space in accordance with the Registration Convention[799]. However, the Act should have established, at least, the general guidelines for the creation of the registry and the minimum information to be recorded.

6. CONCLUDING REMARKS

The South African Space Affairs Act deals basically with the creation of the Council for Space Affairs. It determines its objects and functions and delimitates its responsibilities and those of the Department of Trade and Industry, which it depends upon. Unlike the Australian, and the Russian Federation acts, the South African Space Affairs Act does not fall within the category of framework law. Thus, many significant aspects, such as the registration of space objects, are not even contemplated in the act. Other aspects, such as the mechanism for the reallocation of risks and liability, are merely referred to future regulations or administrative acts but the Act does not even outline the basic guidelines for their future adoption. This may be attributed to the limited participation of South Africa in activities related to the exploration and use of outer space.

The Minister of Trade and Industry has been entrusted with ample authority to make regulations regarding the licensing procedures, measures to protect the national interests, safety measures and minimum safety standards concerning any space activity. From a participatory theory perspective, the Act does not contemplate the possibility of procedures allowing those affected by the law to participate in its enactment. Nonetheless, it provides many opportunities for the space industry and the general public to be actively involved in the different aspects of the implementation of space policy, in

---

[798] *Ibid.* article 23.

[799] Online Index of Objects Launched into Outer Space: Notifications from States & Organizations http://www.oosa.unvienna.org/OSOIndex/docsstatidx.html accessed on June 24, 2001.

particular through the participation in the Council for Space Affairs and its committees.

The Act implemented a licensing system to comply with the authorization and supervision obligations assumed at the international level. The Act does not establish with precision the legal grounds for the award of the license and the Council has ample discretion to fill the legal voids of the Act. Nonetheless, it contemplates the possibility to appeal any decision of the Council to the Minister and to resort to a court of law, for which cases the Act envisaged a clear procedure.

The compliance with the continuing supervision principle is carried out through a system of inspections to the licensee's facilities, which again affords the inspectors appointed by the Council with unclearly defined and discretionary rights.

H. JAPAN

1. NATIONAL APPLICATION OF INTERNATIONAL LAW

The treaty making process is vested with the Cabinet, which manages foreign affairs and concludes treaties[800]. In Japan the Cabinet exercises executive power[801] and consists of the Prime Minister, who is its head and representative, and other Ministers of State, as provided for by law[802]. However, it must obtain prior or, depending on circumstances subsequent, approval of the Diet, Japan's parliament[803]. As argued by Iwasawa, the treaty process in Japan is quite straightforward and it is easier to conclude a treaty than to enact a statute or amend the Constitution. The Cabinet simply requests that the Diet approve the treaty by majority vote, which approval, together with its official publication and promulgation, gives it full domestic force[804]. Finally, the Emperor, with the advice and approval of the Cabinet, signs the instruments of ratification[805].

---

[800] Japanese Constitution, article 73.
[801] *Ibid.* article 65.
[802] *Ibid.* article 66.
[803] *Ibid.* article 73.
[804] B. P. Menard, "Yuji Iswasawa, International Law, Human Rights, and Japanese Law: The Impact of International Law on Japanese Law Evidence of Compliance" (2000) 40 *Va. J. Int'l L.* at 768.
[805] Japanese Constitution, article 7.

In Japan international treaties and customary international law[806] override domestic law, including subsequently enacted law[807]. Furthermore, international treaties are generally considered as self-executing and no enacting or implementing legislation is required[808]. However, Iwasawa cautions that, in practice, Japanese courts are rather reluctant to apply international law directly and prefer to deal with the Japanese Constitution than with international law[809].

2. NATURE OF JAPANESE NATIONAL SPACE LAW

Judged against the major spacefarers, Japan's participation in space activities is rather limited[810]. Additionally, compared with the United States, the former Soviet Union and France, Japan's space program began considerably later[811]. Therefore, as will become apparent throughout the following discussion, its domestic legislation is partial, restricted and it does not even intend to cover all possible issues which could need regulation.

Thus, the main purpose of the law concerning NASDA, which was enacted in 1969, is the regulation of Japan's principal space agency now called Japan Aerospace Exploration Agency (JAXA)[812]. It thus contemplates the agency's responsibilities, objectives, governing bodies and its operating budget, as well as the elaboration of a basic plan for space development[813]. It

---

806 J. H. Jackson, "Status of Treaties in Domestic Legal Systems: A Policy Analysis" (1992) 86 *A.J.I.L.* at 310.

807 Japanese Constitution, article 98.

808 Jackson, *supra* note 806 at 310.

809 Menard, *supra* note 804 at 769.

809 Japanese Constitution, article 7.

810 H. Yoshida, "The meaning of Japan's space commercialization efforts", (1992) 8 Sp. Pol. at 325.

811 T. Nomura, "Japan's new long-term vision creating a space age in the new century", (1995) 11 *Sp. Pol.* at 9.

812 Law concerning the National Space Development Agency, Law No. 50 of June 23, 1969, as amended [hereinafter referred to as "NASDA Law"]. The Japan Aerospace Exploration Agency (JAXA) was born on Oct. 1, 2003, merging the Institute of Space and Astronautical Science (ISAS), the National Aerospace Laboratory of Japan (NAL), and the National Space Development Agency of Japan (NASDA). http://www.nasda.go.jp/index-e.html accessed October 10, 2003.

813 The philosophy of the Japanese space plan is defined as follows: "It enables space to be used as the common property of all mankind in order to contribute to the enduring prosperity of all the living on Earth." Japan's main objectives are (i) the promotion of the advanced space science programs, (ii) full implementation of space activities using the Japanese Experiment Module of the International Space Station

also regulates partial aspects of launch activities carried out by the Japanese agency[814].

The law is silent with respect to almost all of the most important regulatory aspects of space activities. In this sense, it does not determine how the Japanese government will comply with the authorization, supervision and registration obligations derived from the international legal scenario.

However, in 1998 after NASDA's development of the H-IIA, a launch vehicle which permits the Japanese agency to offer launches at competitive rates in the international launch market, the law was amended to reflect the regulatory needs for the launching of space objects of private entities and to introduce a simple risk distribution regime[815]. This amendment, which was the first substantive modification to the law since its enactment in 1969, is the result of an extensive work of several governmental departments. Nonetheless, the space industry was not invited to participate in this legislation process[816].

3. RESPONSIBILITY AND LIABILITY

The Law lays down the basic principles of a general and simple risk sharing regime, which allows the Japanese space agency to assume some of the risks of its customers[817]. This regime, which refers most of the issues to regulations or later determination, is silent with respect to first and second party risks.

It prescribes that all satellites to be launched by the Japanese agency must be insured against third party liability. The purpose of this compulsory insurance is to protect potential victims and to permit the space agency to efficiently carry out launching activities[818]. The law does not establish the amount of the insurance, but it refers its determination to the competent Ministers[819]. However, the law clearly establishes that it must be appropriate from the viewpoint of the protection of victims, which is one of the major purposes of the amendment. According to the Japanese agency practice before

---

and (iii) the development and operation of a new space infrastructure. Nomura, *supra* note 811 at 9.

814 NASDA Law, article 1.

815 M. Sato, "The Japanese Legal Framework: Third Party Liability Resulting from NASDA Launch Activities", (1998) 41 *IISL* at 129.

816 *Ibid.* at 129.

817 NASDA Law, article 24-2 (2).

818 Sato, *supra* note 815 at 130.

819 NASDA Law, article 24.2.

the introduction of this amendment, the amount of third party liability insurance was 20 billion yens. [820] Additionally, drawing on the experience which led to the incorporation of the 1988 amendments to the Commercial Space Launch Act in the United States[821], the Law also acknowledges that the Ministers must take into account insurance costs and other factors in the determination of the insurance amount[822].

The Law also authorizes the space governmental agency to assume all of third party liability for damage caused by the launch, except for damage cause by willful misconduct of the user[823], when the Japanese agency would otherwise be jointly liable with its customer[824]. However, the law is drafted in vague and imprecise terms and it is not clear under what circumstances the governmental space agency could provide indemnification. According to Japanese authors, the intention of the law is to offer a similar indemnification as the one foreseen in the United States.[825]

4. CONCLUDING REMARKS

The law concerned the National Space Development Agency established NASDA –currently known as JAXA- and it limited itself to regulating the agency's responsibilities, objectives, governing bodies and its operating budget. Therefore, it does not deal with several regulatory aspects of space activities, such as, authorization, supervision and registration issues.

Nonetheless, it implemented the basic guidelines of a basic third party liability regime for damage arisen from the launch of satellites, where the Japanese agency's customers are obliged to procure liability insurance and the agency is entitled to provide indemnification for all third party risks. This liability regime aims at protecting potential victims while at the same time affording the Japanese agency the possibility of carrying out launching activities under similarly third party liability conditions as in other launch markets.

---

[820] Sato, *supra* note 815 at 130.
[821] Senate Report, *supra* note 475.
[822] NASDA Law, article 24-2 (2).
[823] The user is referred to in the Law as "those related to the Consigned Launch", which includes any person or entity designated by the Agency and the Consignor in accordance with special arrangements as the persons or entities that are related to the Consigned Launch. NASDA Law, article 24-3 (2).
[824] A/AC.105/C.2/L.224 at 6.
[825] Sato, *supra* note 815 at 133.

I. FRANCE

1. NATIONAL APPLICATION OF INTERNATIONAL LAW

The treaty making process in France is vested in the President of the Republic[826]. The President negotiates, signs and ratifies international treaties. These functions, however, may be delegated to the Minister of Foreign Affairs[827]. There are certain treaties, generally those regarding changes in territory, peace treaties, treaties of commerce, and those related to the establishment of international organizations, which need legislative clearance before their ratification or approval[828].

France is enrolled in the monist doctrine[829] and thus international treaties duly ratified or approved prevail over domestic statutes as from the moment of their publication,[830] subject, however, to a reciprocity requirement[831]. France's subscription to the principle of higher normative rank of treaties over statute law reveals the high degree of the French constitution drafters' trust in international institutions and treaty-making processes[832]. Nonetheless, despite the clarity of these constitutional provisions, the French courts only confirmed the full supremacy of treaties and the power of courts to refuse to apply later laws in violation of prior international agreements[833] three decades after the constitutional consecration of the full supremacy principle[834].

2. NATURE OF FRENCH NATIONAL SPACE LAW

French law is composed of a series of scattered contractual, administrative and regional norms and arrangements which apply to different space activities. It has long been held that the French government's active

---

[826] French Constitution (1958) article 52.
[827] *Ibid.* article 52.
[828] *Ibid.* article 53.
[829] P. M. Eisemann & C. Kessedjian, "National Treaty Law and Practice: Francen" in M. Leigh & M. R. Blakeslee, eds., *National Treaty Law and Practice* (Washington, DC: ASIL, 1995) at 13.
[830] French Constitution (1958) article 55.
[831] E. Stein, "International Law in Internal Law: Toward Internationalization of Central-Eastern European Constitutions?" (1994) 88 *A.J.I.L.* at 429.
[832] Jackson, *supra* note 806 at 313.
[833] Cass. Ch mixte, May 24, 1975, Administration des douanes c. Société "Café Jacques Vabre", J.D.I. 1975, at 801.
[834] T. Buergenthal, "Modern Constitutions and Human Rights Treaties" (1997) 36 *Colum. J. Transnat' l L.* at 216.

participation in space activities was considered a guarantee in itself for the authorization and supervision obligations arisen from the *Corpus Juris Spatialis*[835]and that there was no need for the adoption of domestic space law in France. It has long been held that France needed a specific space domestic law, in particular to deal with its increasing private space endeavors.

Nonetheless, in light of the increasing participation of commercial private entities in the exploration and use of outer space in France the government has entrusted its Ministry of Research to appoint and coordinate several working groups to elaborate national space legislation.[836] In this context, the government recently published a report on the changes in Space Law, based on the work in the fields of launch services, Earth observation by satellite, space radiocommunications and navigation by satellite, and ownership and safety systems on space objects[837]. According to the report, the new legal framework for space activities –launching and satellite operations-will include: (i) the creation of a public authorization and monitoring system, which will make authorization obligatory for all launches either from French territory or from a third country by or for a French citizen; (ii) new procedures for the registration of objects launched into outer space, which will take into account the future Unidroit international system; (iii) the application of the concept of public services for certain public space systems creating original applications; (iv) the confirmation of the risk allocation system –examined below- between the government and private space actors; (v) a revitalized intellectual property system; and (vi) an overall tax system for space activities[838]. The report also advocates for strengthening and clarifying the relationship between CNES and other regulatory bodies in the European context[839].

In light of the importance of France's long standing involvement in space activities, and with the view to contrast the French regime to the other countries' domestic norms in the space field, we will briefly examine the norms engineered for the regulation of commercial space launch services rendered by Arianespace.

---

[835] Clerc, *supra* note 225 at 86.

[836]Gerhard & Schrogl, *supra* note 52 at 20.

[837] Study Report: Evolution of Space Law in France, http://www.recherche.gouv.fr/rapport/droitespace/ accessed on November 7, 2003.

[838] *Ibid.*

[839] *Ibid.*

3. RESPONSIBILITY AND LIABILITY

The French government has created a complex system of redistribution of risks and assignment of liability for Arianespace space launch services, which was modeled after NASA's regime[840]. It is composed of a series of declarations, regional agreements and other administrative norms[841].

3.1. First party risks

First party risks are assumed by Arianespace and its customers, by means of reciprocal waivers of liability[842]. These waivers of liability consist of (i) a general assumption of risks by each party, (ii) the assumption of the consequences of those risks, (iii) a consequent waiver of rights to make a claim for liability, (iv) a waiver for the consequences of the losses suffered, and (v) an indemnification or hold harmless provision in case of actions filed despite the waiver[843]. The objectives sought by the reciprocal waivers of

[840] "Launch Services", *supra* note 11 at 126.

[841] J. Hermida, "Risk Management in Arianespace space launch agreements", (2000) XXV *Ann. Air & Sp. L.* at 143 [hereinafter "Arianespace Risk Management"].

[842] They are generally drafted as follows: "Each Party shall bear any and all loss of or damage to property and any bodily harm (including death) and all consequences, whether direct or indirect, of such loss, damage or bodily harm, (including death), and/or of a Launch Mission failure and/or of a Satellite Mission Failure, which it or its Associates may sustain that arises in any way in connection with this Agreement, or the performance of this Agreement. Each Party irrevocably agrees to a no-fault, no subrogation, interparty waiver of liability, and waives the right to make any claims or to initiate any proceedings whether judicial, arbitral, administrative on this account against the other Party or that other Party's Associates for any reason whatsoever. Each Party agrees to bear the financial and any other consequence of such loss, damage or bodily harm (including death), and/or of a Launch Mission failure and/or of a Satellite Mission Failure, which it or its Associates may sustain, without recourse against the other Party or the other Party's Associates. In the event that one or more Associates of a Party shall proceed against the other Party and/or that Party's Associates as a result of such loss, damage or bodily harm (including death), and/or of a Launch Mission failure and/or of a Satellite Mission Failure, the first Party shall indemnify, hold harmless, dispose of any claim, and defend, when not contrary to the governing rules of procedure, any liability and expense, including attorneys' fees, on account of such loss, damage or bodily harm (including death), and/or of a Launch Mission failure and/or of a Satellite Mission Failure and shall pay all expenses and satisfy all judgments and awards which may be incurred or rendered against that other Party and/or its Associates.

[843] The indemnification provision for claims filed despite restriction of waivers of liability stems from the fact that clauses whose object is the exoneration of responsibility in cases of bodily injury are prohibited under French Law. Therefore, in

liability are basically to limit the claims that might arise from a launch, and to eliminate, or at least reduce, the necessity to obtain property and casualty insurance to protect against claims which may otherwise derive from the launch. These objectives are also the same as the ones of the US private sector first party risk- distribution system. Also as in the US system, the reciprocal waivers of liability act as a mechanism for the transfer of first party risks to the customers, thus exempting the launch provider from damage which it causes[844]. This acts as an exclusion of liability, which constitutes an exception to the fault principle of the French civil law.[845]

The scope of reciprocal waivers of liability is quite broad, for they include (i) damage to property, (ii) bodily harm, (iii) death, (iv) all their consequences, (v) Launch Mission failure, and (vi) Satellite Mission Failure. Unlike in the US regime, the waivers of liability used in Arianespace launch services agreements also cover contractual losses. In effect, they include Launch Mission failure, i.e., the impossibility of placing the satellite in the agreed upon orbit due to problems caused by the space vehicle or the launch itself, and Satellite Mission Failure, i.e., risks of causing damage to the satellite which may impede it to attain the intended orbit or operate successfully in it[846].

The reciprocal waivers of liability encompass both Arianespace and its associates and the satellite owner and its associates[847]. This liability-waiver

---

the event that, for example, employees of the customer suffer physical damages or even death they or their heirs could file a claim before the French courts, which would be admitted despite the waiver of liability contained in the launch services agreement. In such a case, the launch carrier could be condemned to pay damages to that employee or their heirs. If so, the carrier could, in turn, recover damages so paid from its customer by invoking the indemnification and hold harmless provision of the agreement. E. Loquin, "La gestion contractuelle des risques de l'exploitation commerciale de l'espace", in P. Kahn, ed., *L'Exploitation commerciale de l'espace: droit positif, droit prospectif*, (Dijon: Litec Credimi, 1992) at 173.

[844] This is so because it is more frequent for a carrier to cause damage to its customer rather than the reverse.

[845] Couston, *supra* note 47 at 245.

[846] *Ibid.*

[847] The term associate is defined as the personnel, the contractors and subcontractors of the launch company and the satellite owner. Therefore, in the event of an accident triggered by a component of the satellite, Arianespace would be precluded from making a claim against its customer contractor or subcontractor that manufactured the part which caused the accident. At the same time, if an accident causes damage to the satellite and the cause of that accident is found to be a device in the Ariane, the satellite owner may not bring a claim against Arianespace's contractor or

scheme is further complemented by obliging each party to the agreement to make its contractors and subcontractors execute reciprocal waivers of liability so that they will also be banned from filing claims in the event of an accident[848].

3.2. Second Party Risks

Second party international liability[849] risks involve Arianespace, the European Space Agency, its member States and the French government. They refer to the possibility of these governmental and supra-governmental entities' being considered launching states and therefore liable pursuant to the Liability Convention.[850] These risks are distributed on a two-layered basis, where Arianespace assumes liability up to approximately 61,000,000 Euros (formerly 400,000,000 French francs) through insurance and the French government bears all liability claims above that level by means of governmental indemnification.

With respect to the first layer, participants in the Production Declaration requested Arianespace to undertake to reimburse the French Government within a ceiling of approximately 61 million Euros per launch[851], the amount of any compensation it may be required to pay in case of damage caused by Ariane launches to third parties. [852] This assumption of liability by

---

subcontractor that produced said device.

[848] This is generally drafted in Arianespace launch services agreement as follows: "Each Party obligates itself to take all necessary and reasonable steps to foreclose claims for loss, damage or bodily harm (including death) by any participant in the launch activity. Each Party shall require its Associates to agree to a no-fault, no subrogation, inter-party waiver of liability and indemnity for loss, damage or bodily harm (including death) its Associates sustain identical to the Parties' undertaking under this Article ... of the Agreement..."

[849] "Space Risk Management" *supra* note 67 at 6.

[850] "Liability Convention", article 1 *et seq*.

[851] The cap on the reimbursement has been set on a per launch basis. Thus, even if, for example, Ariane carries two payloads in a single launch which causes damages to third parties, Arianespace will still have to reimburse up to approximately 61,000,000 Euros.

[852] Declaration by Certain European Governments Relating to the Ariane Launcher Production Phase signed by states participating in the Ariane production phase, VOL.II-BIS/G02V [hereinafter the "Production Declaraion"], article 3.8. This Declaration, engineered by the French government, entered into force in April 14, 1980 and was signed by states participating in the Ariane production phase. According to the Declaration, the participants decided to entrust an industrial structure, Arianespace, with the execution of the Ariane launcher production phase.

Arianespace is implemented through a reimbursement of costs to the French government for compensation it may have paid in the event of damage caused by Arianespace to third parties if the French government, ESA or its member States were considered launching states and thus held liable for these damages. In this case, Arianespace does not have to pay directly to the victims but has to refund the French government any compensation actually paid by it to third parties or to ESA or its member States if the Agency or its members paid a compensation to the victims of the accident.

The French government also engineered a system for the distribution of risks and assignment of liability for damage caused to the property of the European Space Agency.[853] This system consists of a partial assumption of risks by the launch carrier and a governmental indemnification granted by the French state. The Production Declaration is silent as to the floor and ceiling of this governmental indemnification. However, a thorough analysis of the Declaration indicates that since Arianespace is obliged to face all claims up to approximately 61,000,000 Euros through insurance, the indemnification granted by the French government operates in practice as from that level upwards[854]. Additionally, since there is no cap it may be concluded that the French government has volunteered to indemnify the maximum possible loss, i.e., all claims which may arise from the launch regardless of the aggregate amount and their likelihood of occurrence. Therefore, Arianespace assumes liability for what is considered maximum probable loss, i.e., approximately 61,000,000 Euros, and the government assumes the potential but extremely unlikely maximum possible loss[855].

The objective of this production phase was to meet the launch requirements of the world market subject only to the following conditions: (i) that it should be carried out for peaceful purposes in conformity with the ESA Convention and the Outer Space Treaty, and (ii) that ESA and the participant states should have priority over third party customers. Arianespace was thus assigned the manufacture, marketing and launch of the Ariane launchers.

[853] Declaration by Certain European Governments Relating to the Ariane Launcher Production Phase signed by states participating in the Ariane production phase, VOL.II-BIS/G02V, article 4.1 [hereinafter the "Convention"].

[854] "Arianespace Risk Management", supra note 841 at 143.

[855] The French government assumes the liability of ESA's member states, ESA itself and Arianespace. In order to distribute this category of second party risks, the ESA and Arianespace signed a Convention on May 15, 1981 - extended on September 24, 1992 - which aimed at putting into practice the principles of the Production Declaration. Convention between ESA and Arianespace signed May 15, 1981, ESA/C(81)11

3.3. Third Party Risks

Third party risks are distributed in Arianespace launch services agreements on a two-layered basis. In the first level, Arianespace requires the customer to assume the risks up to the amount of approximately 61 million Euros through insurance taken by Arianespace and paid for by the customer[856]. In the second level, the French government provides full indemnification to Arianespace above approximately 61 million Euros[857].

4. IMPLEMENTATION OF THE AUTHORIZATION AND SUPERVISION PRINCIPLE

The current legal scenario for the implementation of France's obligation to authorize and continuously supervise activities in outer space consists of a set of contractual, administrative and regional measures applied to each French – European- program, such as Starsem, Eutelsat and Ariane, which emphasizes government and –to a lesser extent- agency participation[858].

---

[856] This clause generally reads as follows: "Arianespace shall, for the Launch, take out an occurrence basis type insurance policy at Customer's cost to protect itself and Customer against liability for property and bodily harm which Third Parties may sustain and which are caused by the Combined Space Vehicle or part thereof. In said insurance policy the natural and corporate bodies hereafter shall be named as assured: 1. The government of France. 2. The Centre National d'Études Spatiales "C.N.E.S." and any launching state as per Convention of March 29, 1972 related to the international liability damage caused by spacecraft. 3. The auxiliaries of any kind, whom Arianespace and/or the C.N.E.S. would call for in view of the preparation and execution of the launching operations. 4. The European Space Agency "E.S.A.", but only in its capacity as owner of certain facility and/or outfits located in the Centre Spatial Guyanais in Kourou and made available to Arianespace and/or C.N.E.S. for the purpose of the preparation and the execution of the launches. 5. The firms, who have participated in the design and/or in the execution and/or who have provided the components of the Launch Vehicle, of its support equipment including propellants and other products either liquid or gaseous necessary for the functioning of the said Launch Vehicle, their contractors, subcontractors and suppliers. 6. Customer and Third Party Customer(s) of Arianespace on whose behalf Arianespace executes the launch services as well as their co-contractors and subcontractors. 7. When they act in the scope of their activities, the Officers and Directors, the legal representatives, the Managing Director, the employees, agents, as well as the interim staff employed by Arianespace or by the Assured mentioned in hereabove Paragraph 1 to 6 (included).

[857] Governmental indemnification constitutes a fundamental risk-sharing instrument aimed at protecting Arianespace's customers for claims above the level of insurance. Through this indemnification the customer is relieved from the risks of having to face claims above approximately 61,000,000 Euros.

[858] P. Clerc, "Current French Plans for a National Legal Framework for Space

The regulations governing Arianespace have been adopted at different stages according to the evolution of this program. First, led by France, several European states signed the "Arrangement between certain European Governments and the ESRO concerning the Execution of the Ariane Launcher Programme" on September 21, 1973[859]. Under the terms of this agreement, the participating States committed themselves to carry out the first phase of the Ariane program, which had two main objectives. The first one was to give Europe a capability on its own at the beginning of the 1980's for placing in orbit geostationary satellites developed within the framework of the Organization of the European states. The second objective was to define the launcher and to organize its production in such a way so as to achieve an economically competitive production cost. The program comprised a second phase, which would have as its objective the production of the launcher, whose legal structure and terms would be decided at a later stage[860].

The Organization, acting on behalf of the participants, was the owner of the elements of the Ariane launcher, of the facilities and equipment acquired for its development, and of the launching facilities produced within the framework of the program. Additionally, participants that owned facilities that could be used for the purposes of the Ariane program undertook to make them available on financial conditions limited to marginal cost reimbursement[861]. These elements, facilities, and equipment had to be made available to the participants acting in the framework of their own program or of a program of the Organization, insofar as this did not interfere with their use for the purposes of the Ariane program[862].

The French space agency, *Centre National d'Etudes Spatiales* (CNES), was entrusted with the execution of the program. It was further stipulated that it should enter into an agreement with ESRO –later the European Space Agency- in order to define the provisions to govern the relationship between both organizations regarding the Ariane Project. This agreement, called the Agreement Between ESRO and CNES (France)

Activities", (2001) 44 *IISL* at 153 [hereinafter "French Plans"]. In Clerc's own words "France fulfills its national and international responsibility with Space activities under a contractual and administrative legal regime established program by program."

859 Arrangement between certain European Governments and the ESRO concerning the Execution of the Ariane Launcher Programme" signed 21 September, 1973, ESRO C(73)41.

860 "Space Risk Management", *supra* note 67 at 32.

861 "Arianespace Risk Management", *supra* note 841 at 143.

862 See "Space Risk Management *supra* note 67 at 33.

Concerning the Execution of the Ariane Launcher Program[863], was signed on February 7, 1974 and went into effect retroactively on December 28, 1973. This legal instrument sets the functions of CNES, among which the following stand out: (i) the technical and financial management of the Ariane Launcher program, (ii) the definition and implementation of the organization of the industry in accordance with the geographical distribution of work, and (iii) the selection of the industrial contractors in charge of developing the different aspects of the project[864].

After the creation of the European Space Agency, the Ariane program was transferred to this agency, and it was registered within the legal framework of the ESA optional programs. The main function of ESA with respect to the Ariane program was to control the role of CNES and the contractors' work[865].

The French government engineered the legal framework of the Ariane production phase and prepared a Declaration to govern the aspects of the future organization of the production of the Ariane and the commercialization of its launch services. This Declaration - considered an act of international law[866]- is entitled Declaration of Production and entered into force in April 14, 1980[867]. Under to the Declaration, the participants entrusted an industrial

---

[863] Agreement Between ESRO and CNES (France) Concerning the Execution of the Ariane Launcher Program signed 7 February, 1974.

[864] Regarding the distribution of the contracts related to the different phases of the Ariane program, the policy of just return (*"juste retour"*) has been followed. This policy has been set in Annex V of the European Space Agency Convention, which regulates ESA's industrial policy and elaborates on the general characteristics outlined in article VII of the Convention. According to the just return principle, preference for the award of contracts must be granted to the member states participating in a program in proportion to their contributions. For that purpose, a return coefficient is stipulated for each state. This coefficient is defined as the ratio between the percentage share of all contracts awarded among all member states and each state's total percentage contributions.

[865]Vitt, *supra* note 53. *Aérospatiale* partially merged with Alcatel and joined Matra to form the Aerospatiale Matra group in 1999. The election of CNES as the only program director intended to avoid the dispersion of industrial efforts known in the European past. CNES, in turn, had to award the direction of the industrial work to a French corporation formerly called Aérospatiale, which had to work with French contractors and with contractors of other countries.

[866]M. Bourély, "La Production du Lanceur Ariane", (1981) VI *Ann. Air & Sp. L.* at 293.

[867] Declaration of Production. At its expiration the Declaration was renewed.

structure, Arianespace,[868] with the execution of the Ariane launcher production phase and set out the objectives, rights and obligations of the all the parties concerned with the production of the launcher.[869]

Additionally, since the Ariane vehicles are launched from the Guiana Space Centre, located in Kourou, French Guiana[870], the French government and the European Space Agency executed an agreement[871] whereby ESA has

[868] The states participating in the European Space Agency entrusted Arianespace, a corporation created in 1979 for this purpose, the commercialization of launch services. Arianespace is a private enterprise organized under the laws of France, whose main shareholders are CNES, 41 aerospace manufacturers and engineering companies from 12 European countries and 11 Banks. The European Space Agency is charged with controlling the activities of Arianespace through the participation in the decision making process, and has the possibility of issuing a consultative vote in the shareholders meetings and meetings of the board of directors as well as reviewing the documents submitted to the shareholders and directors of Arianespace.

[869] The objective of this production phase was to meet the launch requirements of the world market subject only to the following conditions: (i) that it should be carried out for peaceful purposes in conformity with the ESA Convention and the Outer Space Treaty, and (ii) that ESA and the participant states should have priority over third party customers. Arianespace was thus assigned the manufacture, marketing and launch of the Ariane launchers. The Declaration also sets the basis for the future relationship between ESA and Arianespace. The participants also requested Arianespace to enter into the following commitments: (i) to observe the principle of peaceful purposes as embodied in the ESA Convention and the Outer Space Treaty, (ii) to respect the geographical distribution of industrial work among the participating states from the development and promotion phases, (iii) to have technical and financial responsibility for maintaining in good operational order the assets made available to it, (iv) to restrict the use of the rights and data made available to it to the requirements of the launcher production, (v) to undertake to pay ESA for the use of the CSG and a fee for each sale, (vi) to give priority to ESA and the participating states over third party customers, (vii) to emphasize the European and multicultural character of the development and production of the Ariane launcher, (viii) to reimburse the French Government within a ceiling of approximately 61,000,000 Euros per launch, the amount of any damages it may be required to pay, (ix) to practice a pricing policy taking into account international competition, and (x) to take cognizance of the Declaration through the Board of Directors.

[870] The Guiana Space Center is the property of CNES. "Space Risk Management", *supra* note 67 at 36.

[871] This agreement, which covered the 1975-1980 period, was extended and partially modified by a protocol dated February 6, 1981. The last agreement between the Agency and the French government was executed on November 29, 1993 and will be in effect until December 31, 2003.

the rights to use the Guiana Space Centre facilities and to build and operate the *Ensemble de Lancement Ariane* (ELA), whose property it holds.[872]

Finally, the relations between the French Government and the Agency concerning the Agency's launch site and associated facilities situated within the CSG are part of another agreement signed on May 19, 1976[873]. It determines the relations between the French Government and the Agency and the rights and obligations with regard to the Agency's launch site and associated facilities situated at Kourou and intended for the activities and programs of the Agency. The French government guarantees the Agency and its personnel unrestricted access to the base and unrestricted use of the base for the purposes of the Agency and of its Member States[874].

This labyrinth of declarations, agreements, and norms shows that the French government, acting through CNES -and the European Space Agency, in a lesser extent- exercises an active participation in the development of launch programs. This close involvement in space activities places the government in a unique position to monitor and control each phase of the activities without the need of a formal legislative or administrative regime[875]. Its active participation and involvement act as a guarantee of the legitimacy of the space activities. This system is compatible with the international regime, which refers to each State the manner and procedure for the implementation of the authorization and continuing supervision obligations. However, at present several private endeavors have emerged, which do not want a close association with the government. Therefore, the industry demands the French government to adopt formal and comprehensive national legislation, which sets clear and transparent conditions for the obtainment of the required authorization.

---

872 In turn, the European Space Agency supplies these facilities to Arianespace. M. Bourély, "Space Law and the European Space Agency", *Space Law, Development and Scope*, edited by Jasentuliyana N., Praeger, 1992 at 95.

873 Agreement on ESA's Launch Site and Associated Facilities at Kourou signed May 19, 1976.

874 The operational use of the base requires use of the facilities, equipment and human and material resources of the CSG. The conditions for such use are defined in the CSG Agreement analyzed above.

875 For all other programs governmental control and supervision at an operational level are delegated to CNES and the Air Force Department. "French Plans", *supra* note 855 at 155.

5. IMPLEMENTATION OF THE REGISTRATION OBLIGATIONS

France has a Space Object Registry, which is operated and administered by the Ministry of Foreign Affairs[876]. The registration of space objects is actually carried out by the Ministry of Foreign Affairs based on the information provided by CNES[877].

6. CONCLUDING REMARKS

France has not enacted specific domestic space law. Rather, French law applicable to space activities consists of a series of scattered contractual, administrative and regional norms and arrangements, which have been adopted for each space program as the needs arose. This is due to the fact that the French government has long had an active participation in space activities, and therefore, it did not consider it necessary to enact specific norms to implement its authorization and supervision obligations. Its sole participation was regarded as the necessary authorization and the permanent supervision of space activities mandated by the international space law instruments. However, in light of the increasing participation of commercial private entities the government has entrusted its Ministry of Research to appoint and coordinate several working groups to elaborate specific national space legislation.

The French government has created a complex system of redistribution of risks and assignment of liability for Arianespace space launch services, which is composed of a series of Declarations, regional agreements and other administrative norms. The main feature of this regime is the important degree of assumption of risks by the government as a consequence of the French government's goal to pursue the maintenance of its –and Europe's- leadership in space.

---

[876] *Ibid.* at 155.
[877] *Ibid.* at 155.

J. CANADA

1. NATIONAL APPLICATION OF INTERNATIONAL LAW

Under Canadian law the power to conclude an international treaty is formally exercised by the Governor General as part of the Royal Prerogative, which was confirmed by the Letters Patent of 1947[878]. Therefore, the Federal Executive Branch negotiates and concludes an international treaty even if its subject matter falls within the exclusive competence of a province. Furthermore, the Federal Executive also has the power to ratify an international treaty without the intervention of the legislature.

Following the transformation approach, international treaties, particularly those that involve a change in existing law[879], may only have binding effect on the domestic plane after they have been adopted by the Parliament, whether Federal or provincial in accordance with the constitutional distribution of competence. The Canadian legislature follows both of the methods available for the domestication of an international treaty, i.e., the incorporation of the text of a treaty *in toto,* such as the Civil International Space Station Agreement Implementation Act[880] and the incorporation of the substance of the treaty into a domestic statute, usually followed for treaties whose object deals with issues already existing under Canadian domestic law[881].

Canadian law is not settled with regard to the value of international law which may conflict with domestic law. With respect to customary law, Canada lacks a definitive judicial decision similar to the United Kingdom's Trendtex leading case[882]. As blatantly put by Stephen Toope: "Canadians simply do not know whether or not customary international law forms part of the law of Canada.[883]" A similar void exists with respect to conflicts between treaties that have been transformed into Canadian law and domestic statutes[884]. As to unimplemented treaties, such as most of the International

---

[878] H. M. Kindred, *supra* note 110 at 184.

[879] R. v. Canada Labour Relations Board (1964), 44 D.L.R. (2d) 440 (Man. Q. B.).

[880] *Civil International Space Station Agreement Implementation Act* 1999, c. 35.

[881] However, according to the Canadian Supreme Court in any case the implementation must be manifest. Kindred, *supra* note 110 at 189.

[882] S. J. Toope, "Re Reference by Governor in Council concerning Certain Questions relating to Secession of Quebec from Canada 161 D.L.R. (4th) 385. Supreme Court of Canada, August 20, 1998" (1999) 93 *A.J.I.L.* at 523.

[883] *Ibid.* at 523.

[884] Kindred, *supra* note 110 at 178.

Space Law conventions, i.e., treaties which have been signed and ratified by the Canadian Federal Executive but which have not been transformed into Canadian Law by an act of Parliament or a provincial legislature, recent decisions of the Supreme Court attribute these treaties a persuasive authority for the interpretation of Canadian statutes[885] or even to shape ministerial discretion[886].

2. NATURE OF CANADIAN NATIONAL SPACE LAW

Despite its long standing active participation in the space field, Canada does not have a specific and comprehensive regulatory framework to implement the obligations assumed at the international level and to govern all aspects of its private industry endeavors in outer space.

The only general policy instrument is the Long Term Space Plan, first adopted in 1986, which set the direction for the nation's space endeavors in very broad terms. The Plan builds on the expertise already developed and demonstrated in the field of telecommunications, remote sensing, and space-based manipulators. Today the Long Term Space Plan, which has been reviewed for the third time, contemplates the development of innovative and emerging technologies to meet the needs of Canadians and to promote the growth and competitiveness of Canada's space industry[887]. However, this policy instrument does not provide the foundations for future regulations and does not even establish the mechanisms or forums for the discussion of future space norms. Furthermore, it does not even purport to deal with any regulatory matter and it contents itself with identifying political objectives in the major sectors of the space arena.

Similarly, the proposed policy and regulatory framework elaborated under the auspices of Transport Canada would continue to treat commercial space launch vehicles as aircraft and would confer authority to the Transport Ministry to authorize launch services under oversimplified and inadequate guidelines[888], which fail to address the needs of the Canadian private space industry.

---

[885] K. Knop, "Here and There: International Law in Domestic Courts" (2000) 32 *N.Y.U. J. Int'l L. & Pol.* 512. The persuasive role of international law in the Canadian courts is more frequent in cases where the Canadian Charter of Rights and Freedoms is involved.

[886] Baker v. Canada (Minister of Citizenship and Immigration), [1999] 2 S.C.R. 817.

[887] It also continues the established tradition of international cooperation which has been the mainstay of Canadian accomplishments in space technology.

[888] Overview of Commercial Space,

Therefore, at present space activities do not have a specific and clear framework. Rather, their regulation is dispersed in a series of scattered norms which have not been expressly conceived to deal with the sophisticated and unique features of the space field and in policy documents often written in an imprecise and incomplete fashion. This lack of an effective legal scenario has originated the consequent legal lacunae and acts as a deterrent for the growth of new private space endeavors, in particular from small and medium companies.

3. RESPONSIBILITY AND LIABILITY

3.1. First party risks

The Regulatory and Policy Framework is silent with respect to any risk allocation mechanism between the launch services operator and its customers[889]. Thus, in principle, parties are free to devise and implement any first-party risk sharing regime. However, as in launches carried out in other countries, the extraterritorial effects of commercial space launch laws, such as US laws, may apply to a launch conducted in Canada, and thus the launch services provider will have to comply with these measures[890].

3. 2. Second party risks

Launch facilities in Canada are contemplated to be operated by the private sector. However, no mechanism for the allocation of risks between the government and the launch range operator has been foreseen[891]. It is submitted in this study that the authorization to operate a launch site should be conditioned to the operator's assumption of all -or part of the risks- derived from the launch from its Canadian launch facilities according to specific policy objectives, which the Canadian government should specifically elaborate.

3.3. Third party risks

Following the US model, the Canadian government will require the launch applicant third party liability insurance for death, bodily injury, loss of or damage to property resulting from activities carried out under the

---

http://www.tc.gc.ca/aviation/regserv/rocket/wwwfiles/about.htm accessed on December 14, 2001 [hereinafter "Overview of Commercial Space"].

889 *Ibid.*

890 "Space Risk Management", *supra* note 67 at 126.

891 "Overview of Commercial Space", *supra* note 888.

authorization in connection with any particular launch. As in the US systems, this insurance will be capped at the maximum probable loss. The amount of the maximum probable loss has not yet been defined[892]. Nor are there any parameters or formulas to determine it in the future. However, the Regulatory and Policy Framework states that the insurance requirements should not place the applicant at a competitive disadvantage, which suggests that the government might contemplate cases of high insurance rates in the world market and lessen the insurance requirements in such circumstances[893]. Furthermore, instead of procuring insurance, the applicant may demonstrate financial responsibility in an amount sufficient to compensate the maximum probable loss. This would permit the applicants to resort to self-insurance or any other risk management tool[894].

The insurance and financial responsibility requirements are complemented by indemnification to the government of Canada against loss in the event of claims brought against the government under international treaties where Canada is accountable. The provisions dealing with this indemnification do not establish any limit. Therefore, it could be argued that in case of an accident causing damage to persons and property of another State and where damage claimed under the Liability Convention exceeds the maximum probable loss the Canadian launch services operator would assume all such losses and not just the liability covered by the insurance (or demonstration of financial responsibility) because the launch operator would still have to indemnify even in excess of the maximum probable loss. The limitation of insurance up to the maximum probable loss is incompatible with an uncapped indemnification to the government. It is thus submitted in this study that the indemnification to the Canadian government should be capped to the maximum probable loss, for otherwise the system would be thoroughly inconsistent and incompatible with the government's purpose not to place the launch services provider at a competitive disadvantage.

Additionally, the regulatory project considers the possibility of the Canadian government indemnifying the commercial launch participants in certain cases. Even if this highly important question has been remitted for later studies, the Regulatory and Policy Framework acknowledges that in certain cases, particularly where the probability of an accident is extremely low, the government may find it appropriate to assume the risk through state indemnification[895].

---

892 "Risk Management in Commercial Launches", *supra* note 458 at 13.
893 "Overview of Commercial Space", *supra* note 888.
894 "Space Risk Management", *supra* note 67 at 9.
895 "Overview of Commercial Space", *supra* note 888.

4. IMPLEMENTATION OF THE AUTHORIZATION AND SUPERVISION PRINCIPLE

As arises from the foregoing discussion, most commercial space activities, in particular space launches, are unregulated in Canada. The Ministry of Transport is considered to have primary responsibility to oversee and regulate international launch activities in Canada[896]. This responsibility derives from a factual misconception of launch activities. In effect, space vehicles are deemed to be aircraft under the Canadian Aeronautics Act and thus the Ministry of Transport has assumed limited regulatory functions. In this respect, the regulations establish that: "no person shall launch a rocket [...] except in accordance with an authorization issued by the Minister pursuant to Section 602.44[897]", which in turn provides that "the Minister may issue an authorization [...] where [...] the launch of the rocket is in the public interest and is not likely to affect aviation safety[898]. In practice, Transport Canada only exercises limited controls over space launches, principally to ensure the safety of the airspace which the space vehicle will penetrate, and to inform other aircraft operators of a launch[899].

However, as arises from the quoted regulations, the Ministry of Transport is expected to issue specific full regulations addressing space vehicle requirements under a comprehensive regulatory program[900]. This program will follow the same regulatory principles adopted for commercial aviation activities, i.e., there will be federal government regulatory intervention, international agreements will be fully respected, international cooperation will be encouraged and regulations will be clear and concise[901].

At present, compliance with the authorization requirement of article VI of the Outer Space Treaty takes the form of an authorization to launch granted by Transport Canada, acting through the Canadian Launch Safety Office. The grounds for future issuances of the launch authorization are expected to be based on internationally recognized standards, such as those published by the US Department of Defense, the International Standards Organization and the American Institute of Aeronautics and Astronautics[902].

---

[896] *Ibid.*

[897] Canadian Aviation Regulations, Part VI, Subpart II section 602.43.

[898] *Ibid.* Part VI, Subpart II section 602.44.

[899] "Overview of Commercial Space", *supra* note 888.

[900] *Ibid.*

[901] *Ibid.*

[902] *Ibid.* These authorizations will apply, among others, to launches carried out from Spaceport Canada. Akjuit Aerospace Incorporated, a Canadian private company, built

Basically these standards follow the US model. The Canadian regulatory and policy framework also acknowledges that the regulations will have to take into account the obligations assumed on the international plane and will ensure the safety of people, property and the environment. For the conduct of safety assessments of launch applications the Ministry of Transport will resort to third party expertise. Transport Canada plans to issue single and multiple launch authorizations. The former will be issued for a single mission from a prescribed location carrying a specific payload. The latter will be issued to a launch operator for a series of launches with similar launch vehicles from the same azimuth range. The launch authorization will contain conditions which all applicants will have to comply with. All launch applicants will have to go through an environmental assessment and will have to comply with all measures imposed to mitigate potential harm. Additionally, in order to fulfill the requirements of article VI of the Outer Space Treaty, Canadian authorities will verify that the purpose of the mission and the payload is peaceful[903]. Military missions will be handled by the Department of National Defense.

With respect to the implementation of the continuous supervision obligation, under the Canadian regulatory and policy proposal Transport Canada will develop an ongoing inspection and monitoring program to ensure compliance with established safety regulations, standards and procedures[904].

From a law reform perspective, the policy documents foresee that the whole Canadian society will be consulted and afforded the opportunity to participate in the development of regulations and regulatory programs[905]. Furthermore, Transport Canada acting through the Canadian Launch Safety Office is mandated to develop its policy and regulations in consultation with the commercial launch industry[906].

In order to manage commercial space launch policy and regulations in Canada the Department of Transport entrusted the Canadian Launch Safety Office with the responsibility of developing and administering safety regulations, standards, policies and procedures governing commercial space

---

a commercial launch site in Churchill, Manitoba for the launch of satellites into polar orbits. http://www.spacetoday.org/Rockets/Spaceports/Canada.html accessed October 10, 2003.

[903] *Ibid.* As arises from our above discussion, the actual implementation of this requirement may exceed the prescription of article VI.

[904] *Ibid.*

[905] *Ibid.*

[906] *Ibid.*

launches in Canada. Its obligations will include the issuance of launch authorizations in accordance with those regulations and standards.

5. IMPLEMENTATION OF THE REGISTRATION OBLIGATIONS

Canada has been the first country to inform the Secretary General of the United Nations of the establishment of the registry required under article II of the Registration Convention. On August 24, 1976 the Permanent Mission of Canada to the United States furnished a *note verbale* to the Secretary General where it reported that it would maintain a registry for the recording of objects launched into space[907].

Canada has been fulfilling its obligations arising from the Registration Convention and article VIII of the Outer Space Treaty in a rather informal way. The Department of Foreign Affairs and International Trade transmits the information required under the international treaties to the Secretary General of the United Nations through the Canadian UN desk. The information is actually prepared by the Canadian Space Agency, which in turn sends it to the Foreign Affairs Department for transmittal to the UN.

Under the proposed regulatory and policy framework prepared by Canada the informal mechanism is expected to continue[908]. However, it is submitted that in light of the experience of the other countries as discussed above, the Canadian space industry will necessitate clear rules for the registration of space objects based on the commitments assumed by Canada on the international plane.

6. CONCLUDING REMARKS

Canada lacks a specific and comprehensive regulatory framework to implement the obligations assumed at the international level and to govern all aspects of its private industry endeavors in outer space. Transport Canada elaborated a proposal for a policy and regulatory framework, which would continue to treat commercial space launch vehicles as aircraft and would confer authority to the Transport Ministry to authorize, oversee and regulate international launch activities under oversimplified and inadequate guidelines.

---

[907] ST/SG/SER.E/INF.2.

[908] "Overview of Commercial Space", *supra* note 888.

The Regulatory and Policy Framework does not include a coherent and comprehensive risk allocation policy. It is silent with respect to first party risks and with respect to the relationship between the launch range operator and the Canadian government. With respect to third party risks, the Canadian government will require the launch applicant liability insurance or the demonstration of financial responsibility up to the maximum probable loss. The launch services provider must also grant indemnification to the government for claims where Canada is accountable under international treaties. However, this indemnification does not establish any limit. It is proposed that this indemnification be capped to the maximum probable loss, for otherwise the system would be thoroughly inconsistent and incompatible with the purpose sought by the government of not placing the Canadian launch services provider at a competitive disadvantage. It is further proposed that the Canadian government should elaborate a comprehensive general risk allocation regime to distribute all the risks arising from space activities according to a clear national space policy. For this purpose, a coherent and comprehensive space policy should be elaborated by reaching consensus among all concerned governmental departments, the space industry, and the general public.

Under the proposed framework, compliance with the authorization requirement of article VI of the Outer Space Treaty would take the form of an authorization to launch granted by Transport Canada and the continuous supervision obligation will take the form of an ongoing inspection and to ensure compliance with future safety regulations, standards and procedures.

From a law reform perspective, the policy documents foresee that the whole Canadian society will be consulted and afforded the opportunity to participate in the development of regulations and regulatory programs. Furthermore, Transport Canada acting through the Canadian Launch Safety Office is mandated to develop its policy and regulations in consultation with the commercial launch industry.

Canada has been fulfilling its registration obligations in a rather informal way, where the Foreign Affairs Department transmits the information prepared by the Canadian Space Agency to the Secretary General of the United Nations through the Canadian UN desk. The proposed regulatory and policy framework contemplates the continuation of this informal procedure. However, in light of the experience of the other countries Canada should adopt clear rules for the registration of space objects based on the commitments assumed by Canada on the international plane.

K. HARMONIZATION AND NEW RULES

On the basis of the foregoing discussion, reference can now be made to recent studies on national space legislation which call for harmonization of the domestic laws and for new international norms, which clearly disagree with some of our hypotheses and findings. Our analysis will depart from the postulates of these studies in order to corroborate our considerations. Thus, we will discuss the studies which advocate for the necessity to harmonize authorization, supervision and indemnification, as well as those which propound the adoption of new international rules or new interpretations of existing ones with the alleged purpose of adjusting the international legal framework to the needs of the private space industry.

1. NO NEED FOR HARMONIZATION OF AUTHORIZATION REQUIREMENTS

The conclusions of the Project 2001 Working Group on National Space Legislation, organized by the Institute of Air and Space Law of Cologne University, recommend the harmonization of national space laws in the area of authorization and supervision of space activities[909]. The reason behind this proposal is that since the core of the national space legislation differs from country to country the space industry may be inclined to direct its activities to the State which provides the most attractive regulatory environment. According to the conclusions of this study, this would lead to an unfair competition among the States to adopt a -or modify its existing- national legal scenario[910].

Frans von der Dunk, in his doctoral thesis submitted to the University of Leiden, also postulates the harmonization of the European licensing systems since this would be "a substantial factor in preserving the interest and strength of the European space industry as a whole [...], allowing Europe to compete globally in space with competitors such as the United States and Russia".[911]

The basic idea of harmonization underlying these positions is that it will enable the industry to have clear and similar rules in all jurisdictions, for harmonization will avoid unfair competition among States and prevent certain

---

[909] Gerhard & Schrogl, *supra* note 52 at 20.
[910] *Ibid.* at 20.
[911] "Spacescape", *supra* note 8at 304.

States from adopting lower standards to attract most of the endeavors, i.e., the so called flags of convenience.

However, these studies fail to consider the already existing harmonizing foundations, guidelines and faculties arising from the international legal framework and identify a problem, which is not such. In effect, first the existing international laws delimit the scope of the authorization and supervision principle. In this regard, an interpretation of this principle in light of the responsibility provisions of the Outer Space Treaty indicates that State authorization and supervision may not be construed to be at the entire discretion of each State, which has to follow the guiding principles and constraints delineated in the *Corpus Juris Spatialis*.[912] Thus, for example, the legal grounds for granting or rejecting the authorization to embark on national activities in outer space is the adherence –or lack of adherence, respectively- to the provisions of the Outer Space Treaty, which are –obviously- identical, for all States. The same applies to supervision. Thus, the international scenario clearly imposes certain harmonizing directives to the States for the implementation of the authorization and supervision principle, which will impede the emergence of the flags of convenience problem. State deviation from this model should be rectified and corrected by urging States to adhere strictly to the international scenario rather than by advocating for the adoption of uniform or harmonized rules. In this respect, the rapporteurs of the Project 2001 Working Group themselves realize the flaw of their argument by recognizing that since authorization and supervision are compulsory international obligations to implement only slight differences might occur[913].

There are, however, certain areas of discretion which we have identified in our examination of the international regime, such as the nature of the entity within the appropriate state to actually grant the authorization or to carry out the supervision, the nature of the legal measures regulating domestic aspects, and the licensing procedure and technical parameters, among others[914].

Differences between laws of different countries within the margins set by the international legal regime are generally legitimate on the basis that they are justified by differences in the substantive concerns such as, the stage of technological development and availability of technology; national preferences, security concerns, state interests, geographical characteristics,

---

[912] See *supra* Chapter I.
[913] Gerhard & Schrogl, *supra* note 52 at 21.
[914] See *supra* Chapter I.

institutional structures, and the ability of different interest groups to organize, formulate and represent policy objectives[915]. Therefore, it would not be advisable to harmonize most of the aspects identified in the Project 2001 Working Group, such as delay of licensing procedure and supervision of space activities[916], since these factors depend largely on the national administrative system where the space authorization regime is inserted and its institutional structures. The same would apply to the technical parameters to ensure public safety. In this respect, it would not be sensible to harmonize, for example, safety precautions for the authorization of launches in all countries, for this would fail to take into account the existing diverse geographical and technological characteristics. For example, safety measures greatly differ – and should continue to differ- in launches made in countries where the launch sites are relatively proximate to major urban conglomerates and those made in countries where the launch sites are thoroughly isolated from human population.

The above discussion makes abundantly clear that the harmonization of the licensing and supervision regimes is not a desirable goal to pursue in the space field. The plethora of alternative licensing and supervision regimes reveals the lack of the States' will for a common approach beyond the parameters established under the international space law instruments. Nevertheless, harmonization could take place in Europe as part of its general integration policy in line of von der Dunk's proposals. This, however, exceeds the scope of the present work.

2. NO NEED FOR HARMONIZATION OF LIABILITY REGIME

The conclusions of the Project 2001 Working Group on National Space Legislation also urge for harmonization of state indemnification. It categorically proclaims that State's indemnification and limitation of liability of the space industry play a significant role in the promotion of commercial space activities and therefore it recommends that the exact composition of this regime should be a primary goal for harmonization prospects[917].

Similarly, but restricted to the European scenario, Frans von der Dunk concludes that: "such topics as parameters for inter party liability arrangements, the question of a ceiling on private liability, and procedural

[915] J. Clift, UNCITRAL and the Goal of Harmonization of Law, Internet Law and Policy Forum, Montreal, Canada, 26-27 July 1999 at 1.
[916] Gerhard & Schrogl, *supra* note 52 at 21.
[917] *Ibid.* at 23.

safeguards for private enterprise in relevant disputes would certainly lend themselves to harmonization efforts[918]."

It is evident from our extensive analysis of all the risk sharing systems that they are a response to the political objectives of the States in the space arena and that they do not necessarily follow the promotion of the space industry as proclaimed by the Project 2001. For example, the United Kingdom reassigns third party liability entirely to the licensee since it does not pursue a policy of promoting space activities or favoring the development of a certain space sector but rather it wants to be thoroughly protected from the financial risk imposed by the international instruments[919]. Under the risk allocation regime engineered by France for the launch services provided by Arianespace, the French government assumes a significant degree of liability as a means of preserving French -and European- leadership in space but not necessarily that of its private industry[920]. Sweden tends to reallocate the risks to the space launch services provider, unless there are special reasons for the State to assume liability[921]. The United State's risk allocation system seeks the promotion of one of the space industry actors -the space launch services provider- in detriment of another -the satellite operator[922]. In Australia, the maximum probable loss is substantially lower than in other countries since the Australian regime is geared to attracting foreign companies to establish a launch facilities industry in Australia[923].

This clearly shows that each country makes use of the risk distribution systems in furtherance of its own space policy. Thus, harmonization of these systems would imply unification of space policies, which is an undesirable and unattainable goal as it implies an overt disregard of political, technological and economic differences of the countries involved and of their space industries.

Von der Dunk's conclusions regarding harmonization of the European indemnification regimes could be seen as a special case, since Europe has long been involved in a comprehensive process of integration, which could be extended to outer space policy. This, again, exceeds the object of the present study.

---

918 "Spacescape", *supra* note 8 at 300.
919 See *supra* note 674.
920 See *supra* note 841.
921 See *supra* note 759.
922 See *supra* note 471.
923 See *supra* note 619.

Thus, in light of the above discussion it is not considered necessary or desirable to harmonize the risk distribution systems of all the countries, since this would imply extrapolating the goal of only one State or a group of States to all others and a clear disregard for the space policy objectives of those other States.

However, due to the consequences derived from international obligations assumed by States, the structure and main elements of the risk allocation regime tend to converge -as this term is understood in Comparative Law-[924] in all jurisdictions, especially in those actively involved in space activities. In effect, legal space risks are allocated among the participants by means of a complex and highly standardized system of reciprocal waivers of liability, indemnification granted by the States, commitments to obtain insurance, limitations of liability, sole contractual remedies in the event of default, and exclusion of liability clauses, among other legal instruments[925]. This, however, does not imply standardization of the content of these systems, which, as discussed above, reflects the unique and distinctive political objectives of each country in the space arena.

3. NO NEED OF NEW INTERNATIONAL RULES

Several prestigious international organizations have recently initiated studies aimed at creating new rules of international law or interpreting existing international norms to adapt them to the new scenario created by the increasing participation of private commercial entities in the pursuit of space activities.

Following the mandate of its 68th Conference, the Space Law Committee of the International Law Association conducted in depth studies geared toward reviewing the Space Treaties so as to establish the need for changes to keep pace with the present commercial context[926]. In connection with these studies, many scholars expressed their view in favor of adopting new substantive or interpretative rules of international law to accommodate the alleged needs of the space industry[927].

COPUOS' Legal Subcommittee also indicated a certain trend to revise existing norms of international law to adapt them to a new private and commercial scenario. For example, it included the review of the status of all

[924] "Modern Approach", *supra* note 51 at 345.
[925] "Space Risk Management", *supra* note 67 at 9.
[926] "ILA Report", *supra* note 82 at 2.
[927] *Ibid.* at 2.

the outer space treaties and conventions as a response to the accumulation of recommendations for defining and extending concepts of space law[928], and the study of the legal concept of launching state as contained in the Liability Convention and the Registration Convention.[929]Furthermore, many of its delegates voiced proposals for reviewing the international treaties[930].

Several authors have also advocated for changing the international legal scenario alleging a need to provide the space industry with a clearer framework. For example, Aldo Cocca recommends the adoption of protocols to revitalize the Outer Space Treaty and the elaboration of new international instruments[931]. Böckstiegel struggled both within the International Law Association and other forums to develop a system of dispute settlement for space activities.[932] Based on the recommendations of international scholars, Eileen Galloway identified an extensive set of issues to reformulate the five legal instruments, which embraces the creation of new international institutions, including a world space agency, the adoption of insurance and liability provisions, and new norms clarifying the relation between national governments and the private sector[933].

The international treaties and conventions dealing with outer space provide a framework which is not altogether responsive to the needs of the private sector, especially with respect to impossibility of making direct claims for compensation under the Liability Convention or the ample mosaic of activities which require authorization and continuous state supervision. However, this scenario is unlikely to be changed in the next few years because of the lack of political will on the part of the main spacefaring nations. Nonetheless, the review of the laws of the States which enacted national space legislation shows that any possible legal vacuum in the international space legal framework has been filled by domestic measures. Thus, for example, States have efficiently adopted a series of mechanisms to deal with the issue of the existence of multiple appropriate States both on a

---

[928] E. Galloway, "Guidelines for the Review and Formulation of Outer Space Treaties", (1998) 41 *IISL* at 245.
[929] Official Records of the General Assembly, Fifty-second session, Supplement No. 20 (A/52/20).
[930] "ILA Report", *supra* note 82 at 14.
[931] A. A. Cocca, "A Way to Complement, Enforce and Improve the Space Treaty and Related International Instruments of Space Law", (1992) 35 *IISL* at 43.
[932] K. H. Böckstiegel, "Developing a System of Dispute Settlement Regarding Space Activities", (1992) 35 *IISL* at 27.
[933] E. Galloway, *supra* note 928 at 249.

case by case basis[934] and through the enactment of specific provisions, such as the extraterritorial effects of the licensing requirements of the Commercial Space Launch Act or the exemption mechanisms contemplated in the Australian Act when a space activity has been licensed by a foreign state, or the exemption certification of the UK Act on Outer Space Activities[935].

Furthermore, the space industry has not advocated for any modifications in the international space legal scenario[936]. Therefore, new international rules or new interpretations of existing ones are clearly not needed, and they even risk imposing new and unnecessary burdens to the space launch industry and its customers[937]. Rather, it is submitted that national legislation should be encouraged in those States which have not yet formulated a comprehensive national legal framework for the regulation of space activities. It is further recommended that this national legislation be based on the common predominant denominators of the national laws of the examined countries, as they provide the proven and efficient instruments to implement the obligations derived from the international legal scenario.

L. CONCLUSIONS

Since under the *Corpus Juris Spatialis* States are free to implement the form of domestic measures to implement the obligations assumed at the international level the nature of domestic law varies from State to State. Except for the United States, most countries which play a preponderant role, or which expect to have an active participation, in space activities have adopted a framework law, such as Russia and Australia. These framework laws are generally wide-ranging in nature, they provide a general regulatory scenario without legislating in detail every single aspect of the space industry, they set the policy basis for future regulation and they refer the regulation of specific aspects of space activities to other agencies. Some acts, such as the

---

[934] For example, under the British Act on Outer Space Activities, the Secretary of State may exempt persons or activities if satisfied that the requirement is not necessary to secure compliance with the international obligations of the United Kingdom. In these cases, an order must be made by statutory instrument, subject to annulment in pursuance of a resolution of either House of Parliament. United Kingdom Act on Outer Space Activities, 1986 Chapter 38, article 3 (3).

[935] See *supra* Chapter II.

[936] Frankle & Steptoe, *supra* note 1 at 10.

[937] *Ibid.* at 10. This study, however, endorses the encouragement to States for the approval of the option included in paragraph 3 of the General Assembly's Resolution 2777 (XXVI) for the decisions of the Claims Commission to be binding.

Russian and the Ukraine laws, lay down the organization of space activities and define the responsibilities of all state agencies involved in the regulation of space activities.

In contrast, US domestic space law consists of a series of laws and regulations which govern specific aspects of different space activities, as well as several non specific norms which have a direct impact on the space industry. This reflects the US early involvement in the exploration and use of outer space, and its reactive approach to deal with space law issues as its needs arose. Nonetheless, US domestic space law is thoroughly comprehensive and covers all possible areas of outer space exploration and use.

The most important space player without domestic legislation specifically focused on space is France. French law applicable to space activities consists of a series of scattered contractual, administrative and regional norms and arrangements which have been adopted for each space program as the needs emerged. The French government, which is currently in the process of elaborating specific national space legislation, has considered that its active participation in space activities fulfilled its obligations under international space law. Similarly, Canada lacks a specific and comprehensive regulatory framework to implement its international obligations. Transport Canada elaborated a proposal for a policy and regulatory framework, which would continue to treat commercial space launch vehicles as aircraft and would confer authority to the Transport Ministry to authorize, oversee and regulate international launch activities under oversimplified and inadequate guidelines.

In those countries with limited participation in space activities, such as South Africa or Sweden, their main legislative instruments do not deal with all significant regulatory aspects. They are generally very succinct and do not even provide the basis for the elaboration of future space policy.

From a law reform and participatory theory perspective, some legislative instruments are the result of extensive negotiations and consultations with local and international interest groups, such as the Australian Act on Space Activities and the US Commercial Space Launch Act. Some States foresee ample participatory mechanisms for the elaboration of regulations, such as the public notice system in the United States, the consultation rounds in Australia and the public consultations in Canada for the development of regulations and regulatory programs. In the United Kingdom the law does not foresee the direct participation of interested parties in the formulation of regulations but these are subject to annulment of either

House of Parliament. The House created a Parliamentary Space Committee to act as a forum of discussion between members of Parliament and the space industry, which has had a major role in the elaboration of space policy. The South African law does not contemplate the possibility of procedures allowing those affected by the law to participate in its enactment but it provides many opportunities for the space industry and the general public to be actively involved in the different aspects of the implementation of space policy, in particular through participation in the Council for Space Affairs and its committees. In other jurisdictions, such as Russia, Ukraine and Sweden, the Law gives marginal participation, if any, to interested individuals and entities in the elaboration of norms and regulations for the exploration and use of outer space.

All major jurisdictions have envisaged a system for the reallocation of state liability arising from the international instruments of Space Law. Even France, which does not have a legislative instrument specifically focused on space, has engineered a risk sharing system for the liability connected with Arianespace launch services. The structure and the elements of these systems present general common features as they all have been modeled after NASA's. However, their actual content differs substantially among States, as these systems are a response to the objectives of the general space policy of each country. Thus, for example, the Arianespace system pursues the maintenance of the French (European) leadership in space, the Australian regime intends to attract foreign companies to a establish launch facilities industry in Australia, and the British and Russian norms intend to liberate the government of any risk by reassigning liability to the licensee but the Russian legislation also allows the government to assume part of the risks. The Unites States regime tends to provide its private sector launch industry with a set of norms that permit it to transfer a significantly high degree of risks to its customers and the government.

All States implemented a licensing system for the authorization of space activities of their non governmental entities. The requirements of these systems, as well as the procedure, differ in all jurisdictions. Some States with limited participation in the space field, such as Sweden, merely enunciated the creation of a licensing procedure but did not adopt specific norms to regulate it. In such cases, the legislation either gives ample discretionary powers to the competent authority or refers the actual process and standards for future regulation. In the Russian Federation the implementation of the authorization principle is twofold. First, the government adopted a licensing system for space activities, and second it established a system of certification of space technology. The same applies to Ukraine, which has a dual licensing and registration system of space facilities. The legislation of countries which are not highly involved in the exploration and use of outer space, such as South Africa, tends to be drafted in wide and imprecise terms.

All States which developed norms to govern the award of licenses created a set of safety measures to ensure that the proposed space activities will not pose perils to the public and its property. These technical measures, which are considered the cornerstone of all licensing procedures, vary from country to country. Generally speaking, they require the verification of all major technical aspects related to the launch. For example, in the United States, the safety review, which is the most important stage of the license procedure, constitutes a mechanism whereby the competent authority examines the launch site, the quality procedures, the capacity of the personnel and the launch vehicle equipment in order to ensure that the launch will not endanger the public safety of the United States.

In some jurisdictions, notably the United States and the United Kingdom, satellite telecommunications are governed by a complex and sophisticated set of rules and are licensed under specific procedures.

The continuing supervision obligation has been implemented mainly through the appointment of observers or delegates at the premises of the non governmental entity that conducts space activities, such as in the domestic laws of the United States, Russia, Australia, and South Africa, among others, and through the possibility of suspending or revoking a license already issued in the event of the licensee's non compliance with conditions or obligations or in the event of extraordinary circumstances which may jeopardize public health and safety as is contemplated in the national legislation of the United States, Russia, South Africa, Sweden, and Australia.

All major space jurisdictions adopted simple and straightforward measures to comply with the registration obligations derived from the Outer Space Treaty and the Registration Convention. In most jurisdictions the information to be recorded follows the standards contained in article IV of the Registration Convention. The United Kingdom adopted a double register system of space objects composed of a main registry and a supplementary register. The former records all space objects except for those whose launch has been procured by a UK satellite supplier, but which appear on the registry of another State. Under Ukrainian law, space facilities, which includes not only space objects but also ground infrastructure, must be recorded in the State Register of Space Facilities. Canada follows a very informal procedure where the Foreign Affairs Department transmits the information prepared by the Canadian Space Agency to the Secretary General of the United Nations through the Canadian UN desk.

Recent studies, which advocate for harmonization of the licensing system in the space arena, fail to consider the already existing harmonizing guidelines, faculties and constraints imposed by the international legal framework. Differences between laws of different countries within the margins set by the international legal regime are generally legitimate on the basis that they are justified by differences in the substantive concerns such as, the stage of technological development and availability of technology; national preferences, geographical characteristics, institutional structures, and the ability of different interest groups to organize, formulate and represent policy objectives. Similarly, despite the conclusions of several recent studies recommending harmonization of national space laws in the area of state indemnification, it is not desirable to harmonize the risk sharing systems, for they reflect the diverse political objectives of the States in the space arena. In this case, harmonization would imply unification of space policies and the consequent elimination of national space objectives. However, due to the consequences derived from international obligations assumed by States, the structure and main elements of the risk allocation regime tend to converge –as this term is understood in Comparative Law- in all jurisdictions, especially in those actively involved in space activities.

Finally, the experience of States which enacted national space legislation shows that new international rules or new interpretations of existing ones are not needed to fill perceived legal vacuums in the international space legal framework, for they risk imposing new and unnecessary burdens to the space industry[938]. Rather, national legislation should be encouraged in those States which have not yet formulated a comprehensive legal framework for the regulation of space activities. These States should profit from the ample experience of States with domestic space law. In this respect, it is recommended that countries without national law base their future legislative instruments on the common predominant denominators of the national laws of the examined countries- as recommended in our proposal[939]- in light of Law Reform and Participatory theories.

---

[938] E. A. Frankle & E. J. Steptoe, *supra* note 1 at 4.

[939] See *infra* Final Chapter.

CHAPTER III

ARGENTINE SPACE LAW AND POLICY

Introduction

We have so far analyzed the foundations arising from the *Corpus Juris Spatialis* and International Law for the adoption of national space legislation and we have examined the actual implementation of domestic measures in the main spacefaring nations. We will now devote to the analysis of the legal framework of outer space activities in Argentina. For this purpose, we will follow the categories of analysis already adopted and used for the analysis of the domestic law in other jurisdictions. Our main hypotheses are that existing Argentine space laws are insufficient to govern current and projected space activities, especially in the area of space transportation, and that Argentine Space Law norms, geared toward hindering the development of space projects and endeavors which do not coincide with national space policy and to providing protection to the dominant service provider in detriment of other existing and potential players, create a legal environment which is hostile to the local private space industry and which impedes the development of many private sector endeavors, particularly in the space transportation field.

In order to contextualize our discussions throughout this chapter we will examine the policy guidelines and objectives outlined in the National Space Plan and the organization of space and satellite telecommunications activities in Argentina, including the functions and structure of the main agencies involved[940].

A. NATIONAL APPLICATION OF INTERNATIONAL LAW

The treaty making process in Argentina is shared between the Executive Federal Branch and the Federal Congress. Thus, the President concludes, i.e.,

[940] J. Hermida, *Norms governing launch services by NASA and commercial US private companies*, (LL.D. Thesis, Catholic University of Cordoba, Doctorate of Laws Thesis 2000) [unpublished]; J. Hermida, "Argentine Space Law and Policy" (1996) XXI-II *Ann. Air & Sp. L* at 177.

negotiates[941], and signs international treaties whereas the Federal Congress approves or rejects treaties concluded with other nations.[942]

The Argentine Constitution is silent with respect to the legislative process to approve treaties and there are no Congressional regulations on this issue. However, it is a well-settled practice in Argentina that treaties are approved by means of laws adopted by the ordinary procedure established in the Constitution for any domestic statute. Thus, the law approving an international treaty must be passed by a majority vote in the House of Deputies and in the Senate. The law is then promulgated by a decree of the Executive Branch and published in the Official Gazette after which the Executive sends the instruments of ratification[943].

According to section 31 of the Argentine Constitution, the Constitution itself, the statutes enacted by the Federal Congress and international treaties are the supreme law of the Nation[944]. Historically, this article has been construed as treating international law on a par with national statutes and consequently a later conflicting legislative enactment overrode an earlier international agreement. However, in 1992 reversing a series of earlier cases the Supreme Court held in Ekmekdjian v. Sofovich[945] that treaties had to be accorded a higher rank than statutes under Argentine law[946]. As has been argued, "Ekmekdjian placed international treaties on a supra-statutory level" in Argentina[947]. The constitutional amendment of 1994 attuned the Constitution to the Ekmekdjian decision and now it can be held that, as a general rule, all treaties are superior to domestic laws.[948] Furthermore, the 1994 amendment conferred constitutional status to a series of Human Rights conventions, specifically listed in the Constitution, which are to be understood

---

941 Under Argentine constitutional law he term "conclude" has been construed to mean "negotiate". J. M. Ruda, "The Role of the Argentine Congress in the Treaty Making Process" in S. A. Riesenfeld and F. M. Abbott (eds.), *Parliamentary Participation in the Making and Operation of Treaties: A Comparative Study* (Dordrecht: Martinus Nijhoff Publishers, 1994) at 180.

942 Argentine Constitution, Article 75.22.

943 Ruda, *supra* note 941 at 181.

944 Argentine Constitution, Article 31.

945 Ekmekdjian v. Sofovich, 315 Fallos 1492, 1511-15 (1992) (Arg.).

946 Buergenthal, *supra* note 834 at 219.

947 J. Koven Levit, "The Constitutionalization of Human Rights in Argentina: Problem or Promise?" (1999) 37 *Colum. J. Transnat'l L.* at 307. The Court further held that international treaties are presumptively self-executing documents as long as they are capable of immediate operation, without additional institutions.

948 N. P. Sagues, "An Introduction and Commentary to the Reform of the Argentine National Constitution" 28 *U. Miami Inter-Am. L. Rev.* at 59.

as complementing the rights and guarantees recognized in the Constitution[949] and left the door open for other treaties and conventions on human rights to attain constitutional hierarchy by means of a specific congressional procedure[950]. Additionally, the 1994 constitutional amendment conferred regional integration treaties and other human rights conventions a special hierarchy status[951].

Potential conflict between international treaties and domestic law under a context of supremacy of international law should be resolved in favor of international treaties. Nonetheless, the Supreme Court cautioned in Hagelin[952] that an international treaty overrides a domestic law "only in the face of a real legal conflict, such that the conflicting laws must be significantly, if not completely, congruent and the underlying purposes behind the laws must be similar. Thus, if a domestic law is designed to deal with a specific problem [...] and an international norm deals more generically with a similar issue, the international norm is not deemed to be in conflict and may not trump the domestic norm"[953].

---

[949] The American Declaration of the Rights and Duties of Man; the Universal Declaration of Human Rights; the American Convention on Human Rights; the International Pact on Economic, Social and Cultural Rights; the International Pact on Civil and Political Rights and its empowering Protocol; the Convention on the Prevention and Punishment of Genocide; the International Convention on the Elimination of all Forms of Racial Discrimination; the Convention on the Elimination of all Forms of Discrimination against Woman; the Convention against Torture and other Cruel, Inhuman or Degrading Treatments or Punishments; the Convention on the Rights of the Child.

[950] A two-thirds vote of all the members of each House is required after their approval by Congress These conventions may only be denounced by the National Executive Power after the approval of two-thirds of all the members of each House. Argentine Constitution, Article 75.22.

[951] Argentine Constitution, Article 75.24. The Constitution establishes a special procedure, which requires an absolute majority of all the members of each House for the approval of these treaties.

[952] Hagelin, Ragnar, CSJN (1993).

[953] Koven Levit, *supra* note 947 at 307.

B. NATURE OF ARGENTINE NATIONAL SPACE LAW

1. CONSTITUTIONAL ISSUES

The Argentine Constitution, whose original text dates back from 1853 and which underwent several amendments throughout its existence[954], does not contain any article relating to outer space[955]. It is thus necessary to elucidate which constitutional provisions govern outer space activities. Neither the courts nor authors have expressly considered this issue. Therefore, we must resort to the analysis of the existing judicial decisions regarding telecommunications and air transport, as these activities are closely linked to space matters and they were not specifically referred to in the Constitution either.[956]

2. JURISDICTION

The first question which arises is whether outer space activities should be considered a federal or a provincial prerogative. This issue is of considerable

---

[954] There have been many proposals to modify Argentina's National Constitution of 1853/60, which is the fourth oldest constitution still in force. Many attempts were made in the Constitutional Convention of 1957 to modify the Constitution; however, those efforts were frustrated by the lack of a quorum. In 1972, the so-called Fundamental Statute, which was dictated by a de facto military government, introduced a series of constitutional amendments. These amendments are no longer in force, however, because of a curious self-destruction provision in Article 4 of the Fundamental Statute. After Argentina returned to democracy in 1983, the reform process was stimulated by various proposals presented in Congress by legislators from diverse political parties. Sagues, *supra* note 948 at 43.

[955] The last amendment took place in 1994 and it did not include any reference to outer space issues. The 1994 constitutional reforms are reflections of the myriad experiences which Argentina and Latin America have had with totalitarian rule. The reforms are a partial attempt to address this history and to resolve the tension between either subordinating or wholly incorporating US and other international precepts and principles. The reforms are a careful synthesis attempting to construct an administration of justice whose probity can withstand the unique challenges of Argentina's complex and sometimes troubled political history. S. N. Vittadini Andrés, "First Amendment Influence in Argentine Republic Law and Jurisprudence" (1999) 4 *Comm. L. & Pol'y* at 149.

[956] This does not presuppose that space and aviation issues are necessarily analog, but the comparison is adequate for the purpose of determining the applicable fundamental provisions of the Argentine Constitution. In the latest amendment of the Constitution in 1994 the text incorporated some articles referring to telecommunications, which confirmed the approached followed so far.

importance because it will determine the appropriate organisms in charge of enacting domestic space norms.

In general, in Argentina the federal government enjoys limited powers while those of the provinces are more plenary[957]. The role of the federal government is restricted to establishing minimum standards regarding economic matters, and to taking responsibility for interstate commerce, international agreements and foreign trade[958]. Furthermore, under the Argentine Constitution the provinces reserved to themselves all the powers not expressly delegated to the federal government in the Constitution or in special pacts at the time of their incorporation[959]. Neither the Constitution nor the pacts refer to outer space issues. So, a literal interpretation could lead to the conclusion that the domestic regulation of space activities should be considered a provincial prerogative. This argument, however, is not compatible with the judicial interpretation of similar issues, especially in the telecommunications field. It also ignores the faculties of the Federal Congress with regard to the implementation of international treaties.

In effect, the Supreme Court has long held that the federal government has jurisdiction over telecommunications services.[960] In *Transradio Internacional v. Provincia de Buenos Aires*[961], the Supreme Court resorted to the interstate commerce clause to attribute federal jurisdiction over telecommunications[962]. It held that both the international and interprovincial communications systems are an inseparable allied of commerce and thus their regulation pertains exclusively to the Federal Congress. In a later decision, the Supreme Court used similar arguments to establish the competence of the federal authorities over broadcasting services[963]. Thus, federal jurisdiction preempted local jurisdiction in all telecommunications matters. A series of decisions of the federal courts in air transportation matters at the dawn of the aviation industry in Argentina also conferred federal jurisdiction over air

[957] Nolon, *supra* note 48 at 685.
[958] *Ibid.* at 685.
[959] Argentine Constitution, article 121.
[960] The first decision was issued in 1874. The Supreme Court held that according to the existing acts at that time the federal government had exclusive authority over telegraphic services and that telephony could be assimilated to telegraphy. Therefore, federal jurisdiction preempted local jurisdiction in all telephony issues. Supreme Court of Justice Decision 179, p. 246 and Decisions 184 p., 280.
[961] Supreme Court of Justice "Transradio Internacional v. Provincia de Buenos Aires", Decision 269, p. 92.
[962] Argentine Constitution, article 75.13.
[963] Supreme Court of Justice "Roca Hnos. v. Provincia de Santa Fe", Decision 188, p. 24.

transportation issues on similar grounds[964]. From a constitutional law perspective, the nature of telecommunications and air transportation issues does not differ substantially from outer space activities in general. They are high technology activities which constitute an essential aspect of modern commerce. Thus, it is submitted that these Supreme Court decisions could also be extended to all other outer space activities.

Furthermore, the interstate commerce clause mandates the federal government to regulate interprovincial and international trade[965] and at the same time it prohibits the provinces to "enact laws dealing with commerce, inland or foreign navigation.[966]" Almost by definition outer space activities do not recognize interstate or international boundaries[967]. In Argentina outer space activities are envisioned, produced and developed in different provinces and are carried out in cooperation with foreign states[968]. Thus, it is submitted that the interstate commerce clause also applies to the regulation of all outer space activities, and therefore the regulation of outer space activities is a matter of federal jurisdiction. Consequently, only the Federal Congress may enact legislation concerning the promotion, introduction and establishment of the space industry, the construction of space launch sites, the exploration of outer space and the enactment of laws protecting space activities and the space industry through temporary grants of privileges and stimulating rewards[969] and the promotion of scientific and technological research and development related to Outer Space[970]. Additionally, all disputes related to

---

964 Horowitz v. Cruzeiro do Sul, Cámara Comercial de la Capital, 9-VI-1948, Gerardo Ramón y Cía SRL v. Panair do Brazil, Cámara Comercial de la Capital, Sala A, 25-II-1965.

965 Argentine Constitution, article 75.13.

966 Argentine Constitution article 126.

967 "Commercial Space", *supra* note 7 at 20.

968 S. C. Negro, Cooperación Espacial Comunitaria. Régimen Jurídico. Exploración y Explotación del Espacio (Buenos Aires: Ediciones Ciudad Argentina, 1997) at 72.

969 Argentine Constitution, Article 75. 18. It reads as follows: "Congress is empowered: [...] To provide for the prosperity of the country, for the advance and welfare of all the provinces, and for the progress of education, drawing up general and university educational plans, and promoting industry, immigration, the construction of railways and navigable canals, the colonization of government- owned lands, the introduction and establishment of new industries, the imports of foreign capital, and the exploration of inland rivers, through laws protecting these aims and through temporary grants of privileges and stimulating rewards.

970 Argentine Constitution, Article 75. 19. It reads as follows: "To provide everything relevant to human development, economic progress with social justice, the growth of the national economy, the creation of jobs, the professional training of workers, the

outer space activities may only be heard by federal courts of justice[971] and the Supreme Court will have appellate jurisdiction, according to regulations and exceptions prescribed by Congress[972].

The Argentine Constitution also mandates the Federal Congress to approve or reject treaties concluded with other nations.[973] Furthermore, article 27 confirms that the Federal government has not only the power but also the obligation to enter into international treaties to strengthen relations with foreign nations[974]. Therefore the implementation of the obligations which arise from those treaties should also be a federal matter. Thus, in addition to – and regardless of- the application of the interstate commerce clause it is held that domestic space laws which are enacted as a response to the obligations contained in the international instruments of space law are the exclusive prerogative of the Federal government.

The existing norms dealing with outer space activities, while not directly mentioning any of these interpretations, support the conclusion that the regulation of outer space activities belongs to the federal domain[975].

3. CONSTITUTIONAL PRINCIPLES

There are certain constitutional principles, rights and guarantees which are specific in nature and apply to those activities expressly referred to in the

---

defense of the currency value, the scientific and technological research and development, their overall diffusion and beneficial use.

[971] Argentine Constitution, Article 116. It reads as follows: "The Supreme Court and the lower courts of the Nation are empowered to hear and decide all cases arising under the Constitution and the laws of the Nation, with the exception made in Section 75, subsection 12, and under the treaties made with foreign nations; all cases concerning ambassadors, public ministers and foreign consuls; cases related to admiralty and maritime jurisdiction; matters in which the Nation shall be a party; actions arising between two or more provinces, between one province and the inhabitants of another province, between the inhabitants of different provinces, and between one province or the inhabitants thereof against a foreign state or citizen."

[972] Argentine Constitution, Article 117.

[973] Argentine Constitution, Article 75.22.

[974] Ruda, *supra* note 941 at 179.

[975] However, the Executive Branch and its agencies have usurped congressional faculties. Support for this conclusion is also found in the Bill on Space Activities, which declared that the regulation of space activities is a national legislation matter and that any conflict arisen with respect to space activities shall be heard by the federal courts of the Argentine Republic. Bill on Space Activities, article 44.

Constitution. These include maritime navigation and commerce, livestock, customs, and posts among many others[976].

An analogy of the development of aviation constitutional issues will help to elucidate the specific constitutional norms applying to the regulation of outer space. In this respect, a report prepared by the Argentine Aeronautical Association, which reflects the thoughts that inspired the drafters of the Argentine Air Code[977], help to shed light on the discussions that took place with regard to this same problem in the aviation field[978]. It was argued that there existed two possibilities to deal with the determination of these constitutional rights. The first alternative implied the amendment of the Constitution to incorporate specific norms dealing with aviation matters, such as federal jurisdiction, the no imposition of transit duties and the adoption of an Air Code[979]. However, this alternative implied that without such amendment these issues would not have a constitutional status and could be subordinated to governmental discretion at least until an amendment incorporated them to the Constitution. The other alternative entailed interpreting that the constitutional articles dealing with maritime commerce also apply to commercial aviation activities[980]. A series of judicial precedents support the argument that that aviation matters should be treated on a par with maritime commerce for Argentine Constitutional Law purposes[981].

Given the fact that there is no reason to treat aviation and space matters differently from an Argentine Constitutional Law perspective since, like the maritime issues, both are activities with a strong international aspect, it is submitted that the same construction may be extended to the space field for Argentine Constitutional Law purposes. Thus, for example, the transit of satellites, space launch vehicles and their components, whether manufactured in Argentina or in a foreign country, are free from the so called transit duties and from any other duties imposed on them by reason of their passing through the territory. This applies when satellites and space vehicles are carried from

---

[976] Argentine Constitution, articles 49, 11 and 26.
[977] F.Videla Escalada, *Derecho Aeronáutico* (Buenos Aires: V.P. de Zavalia, 1969) at 1027.
[978] Asociación Aeronáutica Argentina, *Informe sobre Política Aérea en Materia de Jurisdicción Nacional y Provincial* (AAA: Buenos Aires, 1985) at 9.
[979] *Ibid.* at 9.
[980] Videla Escalada, *supra* note 977 at 1027.
[981] Itoiz de Anaorena et al. v. FAMA. Cámara Comercial de la Capital, 27'VIII-1948, Horowitz v. Cruzeiro do Sul, Cámara Comercial de la Capital, 9-VI-1948.

one province to another for launching or testing purposes[982]. Also satellites or space launch vehicles –and their component parts- flying through the territory of one province may not be required to seek any authorization from the government of that province[983].

3.1. General constitutional prescriptions and rights

There are other important provisions in the Constitution which have a general scope and there is thus no doubt that they apply to outer space activities. Therefore, for example, satellites and space launch vehicles have to respect the constitutional prescriptions regarding the protection of the environment, which were introduced in the 1994 amendment in response to the accords reached at the Earth Summit in Rio de Janeiro in 1992. These include the obligation to respect the right of all residents to a healthy environment[984] and the prohibition to enter radioactive debris in the national territory[985]. National space activities must also satisfy present productive needs without compromising the needs of future generations[986].

The Argentine Congress is now faced with the obligation to enact federal legislation to regulate these constitutional rights and obligations. Until the adoption of a law, the environmental responsibilities foreseen in the Constitution are not clearly defined and may be difficult to comply with in practice. Nonetheless, there are few environmental provisions in the Constitution, such as the prohibition to introduce radioactive debris in the national territory, which all space activities must respect, regardless of the existence of a statute regulating this issue and regardless of the absence of such prohibition in an international Space Law treaty.

---

[982] Argentine Constitution, article 11. This reads as follows: "Goods of national or foreign production or manufacture, as well as livestock of all kinds, that may pass through the territory of one province to another, shall be free from the so called transit duties, the same as the carriages, vessels or beasts in or on which they are transported; and no other duty, whatever its name may be, shall be imposed on them by reason of their passing through the territory."

[983] Argentine Constitution, article 12. This article reads as follows: "Vessels sailing from one province to another shall not be bound to enter, anchor, or pay transit duties; and no preference shall be granted in any case to any port in respect of another, by means of trading laws or regulations."

[984] Nolon, *supra* note 48 at 688.

[985] Argentine Constitution, article 41.

[986] It has also adopted the international definition of sustainable development as a means of protecting this environmental right. Nolon, *supra* note 48 at 689.

Likewise, all the guarantees and rights that protect all Argentine inhabitants and entities extend to those engaged in the exploration and use of outer space. These guarantees would include, for example, the free circulation of goods of national production[987], the protection of property[988], and the rights to work and perform any lawful industry[989], to navigate and trade[990], and to enter, remain in, travel through, and leave the Argentine territory[991], among many others.

4. LACK OF CONSTITUTIONAL AUTHORITY

Based on the foregoing discussion, only the Federal Congress may enact legislation concerning the regulation of most aspects of space activities and it has also exclusive authority to implement the obligations arising from international treaties. However, the Federal Congress has played a marginal role in the adoption of national space law and most of the existing norms emanate from Executive Branch agencies. In effect, the main norms of the national Space Law framework emanate from divisions of the Executive Branch, in particular from the National Commission on Space Activities (CONAE) and the National Communications Commission (CNC). The Congress' role has been practically limited to approving the international space treaties and conventions and even in these cases there has been practically no debate and the treaties have been approved without an analysis of their main provisions and their implication for the Argentine Republic.[992] Thus, there are no domestic congressional interpretations of the treaties which Argentina is a party to[993]. So far, Argentina has ratified the Outer Space Treaty, the Liability Convention, the Rescue and Return Agreement and the Registration Convention[994].

---

[987] Argentine Constitution, article 10.

[988] *Ibid.* article 17.

[989] *Ibid.* article 14.

[990] *Ibid.* article 14.

[991] *Ibid.* article 14.

[992] "Argentine Space Law and Policy", *supra* note 421 at 177.

[993] Furthermore, the decision not to ratify the Moon Agreement was not actually taken by Congress but by the Executive Branch that did not introduce a bill to Congress so that it would debate it.

[994] On December 4, 1968 Argentina ratified the Outer Space Treaty shortly after its adoption by the UN Assembly. By means of Act No. 23335 of July 30, 1986 Argentina ratified the Liability Convention. After a lobbying movement initiated by the Integration Foundation, a civil organization leaded by a former Minister of Justice with political aspirations to be considered for presidential candidate, through Act 24158 the Argentine Congress ratified the Registration Convention on September 30, 1992. Despite the fact that Argentina has taken the initiative to elaborate the Moon

Even the legal framework of the governmental agencies which are entrusted with outer space responsibilities has been structured by the Executive Branch through presidential decrees rather than by Congress.[995] It is thus necessary to scrutinize the legal basis of the norms issued by these agencies, i.e., the legal grounds on which they were adopted, to elucidate whether they have been enacted in conformity with the national constitutional order. For this purpose, first it is necessary to examine the functions of CONAE, which will permit us to contextualize the present discussion.

5. FUNCTIONS

Pursuant to Decree No. 995/91, CONAE is the only competent state agency to design, execute, control and administer outer space projects and endeavors[996]. The presidential directive required CONAE to draft the National Space Plan for the use and exploitation of space science and technology with peaceful purposes, and to execute such plan once approved by the President. Additionally, CONAE is entrusted with the following functions: (i) centralize, organize, administer and execute the National Space Plan, (ii) to carry out space research activities, (iii) to execute and develop national space programs, (iv) to transfer space technology to state entities and to the private sector under a license, (v) to train researchers, scientists, technicians and professionals, (vi) to enter into cooperative agreements with public and private entities, (vii) to coordinate all the activities of the National Space System, which includes all the private and public institutions, which directly or indirectly carry out space activities, and (viii) to promote cooperation agreements with other States and foreign entities[997].

As can clearly be seen from the above enumeration of functions, CONAE does not have any regulatory powers. Nonetheless, it has been argued that CONAE is a regulatory agency and that as such it has full authority to adopt resolutions with a wide reaching scope[998]. Furthermore, the

---

Agreement in the United Nations thanks to its then representative, Aldo Armando Cocca, Argentina has not yet ratified this Agreement.

[995] Argentine Decree 995/91 creating CONAE, Argentine Decree 1185/90 creating CNC.

[996] Argentine Decree 995/91, article 2.

[997] The terms "outer space programs" and "outer space endeavors" have not been defined and CONAE has understood that they include satellite telecommunications. Thus, as analyzed below, this has led to tensions between CONAE and CNC, which resulted in contradictory legal measures and political tension.

[998] "Recopilación y análisis", *supra* note 247 at 6.

resolutions adopted by CONAE justify CONAE's own power to adopt these norms in its mandate to "coordinate all the activities of the National Space System, which includes all the private and public institutions, which directly or indirectly carry out space activities", which is contained in CONAE's constitutive decree[999]. It is CONAE's position that in order to comply with its coordination functions, it must centralize all the information related to space activities and enact the necessary norms to achieve this purpose.

This position is erroneous because of the lack of specific authority in CONAE's constitutive norms, the exclusive prerogative of the Federal Congress to implement the obligations arising from an international treaty, and an accurate interpretation of the extent and nature of the coordination functions in light of the role assigned to CONAE under the National Space Plan. First, as is clear from the analysis of CONAE's functions as contained in CONAE's constitutive instruments, it lacks any regulatory power. In effect, nowhere in the norm creating CONAE is there any delegation of regulatory powers. This is so, because as discussed above, many of the aspects related to outer space are the exclusive prerogative of the Federal Congress, either on account of the interstate commerce clause or the faculties concerning the adoption of national measures arising from international treaty obligations. Thus, absent an express delegation, CONAE may not issue resolutions aimed at adopting national measures arising from international treaty obligations[1000]. Under Argentine Constitutional Law, the Federal Congress may delegate these faculties to the Executive Branch. It may do so in a specific act or in the case of international treaties it may also delegate these functions in the same act that ratifies the treaty[1001]. As arises from our preceding discussions, Congress has not delegated these functions to the Executive Branch or any of its agencies. Therefore, for example, CONAE's Resolution 330/96 on the disclosure of satellite and related activities which intends to implement the authorization and supervision principle pursuant to the Outer Space Treaty in Argentina should be considered invalid, for CONAE lacked authority to adopt it.

Finally, a thorough interpretation of CONAE's coordination functions in light of the role assumed by CONAE pursuant to the National Space Plan further demonstrates that CONAE has not been entrusted with any regulatory functions. In this sense, Decree No. 995/91 prescribes that CONAE has to

---

[999] Argentine Decree 995/91, article 1.

[1000] All international treaties currently ratified, as well as those that Argentina may ratify in the future, are superior to domestic laws, which may not trump an international norm. Koven Levit, *supra* note 947at 281.

[1001] Ruda, *supra* note 941 at 181.

"centralize, organize, administer and execute the Plan."[1002] It is thus necessary to examine the functions assumed by CONAE under the National Space Plan to determine whether, even in violation of constitutional provisions, it gives CONAE any regulatory functions.

According to the language of the National Space Plan, CONAE's role in space activities should vary according to the economic return of the different space endeavors. Thus, it purports to act as a mere promoter of activities which are transferable to the Argentine commercial private sector, such as satellite voice and data telecommunications, and is engaged in research and development in non-massive activities which have a lower cost-benefit ratio, but which are highly relevant from a social standpoint.[1003] CONAE holds that while maintaining certain features as an executor, it tries to encourage performance by third parties, giving preference to its role as a promoter of technological development instead of being a space service provider or operator[1004]. Additionally, CONAE is directed to promote research and technological development programs, provided that these: "(i) contribute to the creation or consolidation of an infrastructure of data and services by which space activities result to be profitable because of their social or economic impact, (ii) are direct contributors to the technological development involved in the Space Program; and (iii) contribute to the enhancement of the

---

[1002] Argentine Decree 995/91, article 2.

[1003] However, even in this latter case, CONAE claims that it tends to assign as many tasks as possible to the private sector in order both to promote national private space industry and to keep a relatively low budget. Opportunities for the private sector are structured through a series of contracts with institutions, researchers and technologists of the national scientific system and companies, which are grouped under the following categories of programs: (1) propagation and interactions of electromagnetic radiation; (2) physics and chemistry of the atmosphere and the oceans; (3) materials science and solid-state physics; (4) applied mathematics and computation; (5) solar-terrestrial physics and astrophysics; (6) solid ground geophysics; (7) space and microelectronics engineering and instrumentation.; (8) juridical, social, ecological and environmental studies and (9)the economics of space technology and business development.

[1004] As can clearly be seen from the quoted provision of the National Space Plan, the space programs which CONAE may promote are only those which fall within the previous considerations. Therefore, except for remote sensing endeavors which will generally fall under the first proviso, if a space program does not directly contribute to the development of technology related to the National Space Plan and does not enhance education professionals such space endeavor may not be promoted by CONAE. In light of CONAE's peculiar interpretation of its constitutive norms and the National Space Plan, in practice CONAE will intend to abort these projects. National Space Plan at 14.

advanced teaching staff dedicated to the generation of human resources for space activities"[1005].

From the foregoing examination of CONAE's functions and the actual role that CONAE has to play under the National Space Plan it follows that CONAE does not have any regulatory authority. This is so because, as mentioned above, CONAE has not been entrusted with regulatory powers, the Federal Congress has exclusive prerogative to implement the obligations arising from an international treaty, and the role assigned to CONAE under the National Space Plan is that of coordination and not regulation. Therefore, it is submitted that absent a specific delegation of power for the adoption of a specific measure, CONAE may not issue resolutions. It is further recommended that the Federal Congress assume a more active and preponderant role in the formulation of national space law, including the domestic implementation of the international treaty obligations.

6. INADEQUACY AND INCOMPLETENESS OF THE CURRENT NATIONAL SPACE FRAMEWORK.

6.1. Introductory remarks

It has been argued in the literature that Argentina has a fairly comprehensive and adequate national framework for the regulations of outer space activities[1006]. We will examine all the specific norms adopted by Argentine authorities in order to corroborate or refute this premise. For this purpose, the present section will scrutinize the National Space Plan, which sets the policy and guidelines which the subsequent regulations are to be based upon, and will briefly examine the presidential decrees and the resolutions adopted by

[1005]In order to comply with the purposes of its creation and the objectives of the National Space Plan, CONAE has structured its actions and plans around a matrix structure, i.e., a rectangular array of functions arranged in horizontal rows and vertical columns, that includes projects (horizontal rows) and activities (vertical columns). The projects include a set of actions with clearly defined starting and finishing points. Each project has its own schedule, budget and financing. Activities are "actions of a technical or administrative type that are performed either temporarily or regularly, without a definite completion date," whose purpose is to support projects. Both projects and activities are part of missions, which is a concept used for facilitating control of the management of the National Space Plan with no budgetary implications. A mission encompasses all the steps taken with respect to a certain flight to Outer Space. Unlike NASA, there is no defined procedure for individuals or corporations to propose Outer Space missions or projects, which are selected exclusively by CONAE itself. National Space Plan at 15.

[1006] "Recopilación y análisis", *supra* note 247 at 53.

the Executive Branch agencies. Since the National Space Plan has not been the object of much attention in the legal literature, we will examine it in detail so as to provide an adequate framework for our discussion.

6.2. National Space Plan

6.2.a. Premises and methodologies

The analysis of the National Space Plan, in particular its objectives and priorities, will help to determine whether Argentine space policy aims at fostering all sectors of the space industry or whether it privileges certain sectors in detriment of others. Additionally, it will help us deconstruct the political foundations of the existing legal norms.

The National Space Plan prepared by CONAE and approved by a presidential decree[1007] is the main legal and political instrument dealing with Argentine outer space issues. The National Space Plan is based on the premise that Argentina is a space country, i.e., it makes intensive use of products and services derived from outer space activities[1008]. Based on this premise, the National Space Plan tries to define the methods which will permit Argentina to have access to outer space products and services in the future, reaching the conclusion that Argentina will not only be an active consumer but also a producer of these goods and services[1009].

The National Space Plan thus identifies the Space Information Cycle ("SIC") as its conceptual framework. The whole Plan has been conceived and implemented around this concept. The SIC has been defined as the "set of steps encompassing space data generation, transmission, processing and

---

[1007] Argentine Decree 2076/94. It was ratified by the Congress for budgetary purposes without any debate or discussion.

[1008] National Space Plan at 3. The National Space Plan gives the following reasons supporting this pivotal premise, which we literally transcribe: (i) It covers an extensive geographical territory, ranging from the tropic to the pole; (ii) economic activities in Argentina are strongly influenced by extensive primary exports (agricultural, fishing, forestry and mining products, plus hydrocarbons), (iii) its society has a development level that requires everyday use and exchange of detailed and quantitative data on its own structure and economy; (iv) the distinctive distribution of its population demands an intensive use of telecommunications; (v) its large productive areas are vulnerable to natural and anthropogenic catastrophes; and (vi) the regional and international links and commitments undertaken by the nation will oblige it to generate and use goods and services deriving from space science and technology.

[1009] National Space Plan at 3.

use"[1010]. According to the text of the Plan "along this cycle, space activities operate either as use promoters or as suppliers or consumers of both the data itself and the means for its production, transmission, elaboration and storage."[1011]

It arises from the foregoing that satellites are considered as secondary elements whose importance is subordinated to that of the data itself. The rationale of this conception is that society in general may actually benefit from the products of the satellites rather than from the development of satellite technology itself[1012]. Therefore, CONAE's main efforts and resources have been concentrated on the reception and distribution of data –mainly from remote sensing– instead of boosting the Argentine space industry and improving space technology, whose benefits for society would be vaster and everlasting. Under the view adopted in the National Space Plan, the role of space launches and space launch vehicles is even more marginal, for they are considered as mere auxiliary tools of a secondary activity.

6.2.b. Economic analysis of space benefits

From an economic point of view, the National Space Plan conceives space activities as medium and long term investments, thus estimating an internal return rate associated to research and development activities, innovation and related services[1013]. On the basis of these estimations, it concludes that a space program in Argentina is not only potentially sustainable by the local economy but it may also result in significant benefits. It also promulgates that "space activities must be kept closely linked to research and development teams, involving the promotion of these activities as part of their ordinary programs in the understanding that the resources applied for their execution constitute an investment whose return is part of the produce from space activities.[1014]"

Based on the foregoing premises, the National Space Plan identifies the following sectors as the ones which will benefit the most from a direct economic impact in the short and medium terms:

1) Telecommunications.
2) Use of global positioning systems.

[1010] *Ibid.* at 4.
[1011] *Ibid.* at 5.
[1012] "Recopilación y análisis", *supra* note 247 at 54.
[1013] National Space Plan at 6.
[1014] *Ibid.* at 7.

3) Prevention, evaluation and follow-up of natural and anthropogenic disasters.
4) Monitoring and usage of natural resources.
5) Remote detection and control of industrial parameters.
6) Cartography and cadastre.
7) Supervision and quantification of agricultural and forestry production.
8) Fishing, exploitation and surveillance of coastal and oceanic resources.
9) Studies on environmental quality, degradation and contamination.
10) Local and global meteorological studies.
11) Utilization of soil and underground resources.
12) Design of new methods for development management and administration at a regional scale.
13) Global change.

As can be seen from the enumeration above, all of the industrial sectors, except for telecommunications and global positioning systems, refer to sectors where remote sensing activities play a fundamental role. The rationale behind this enumeration is that CONAE privileges remote sensing activities, particularly the reception of data, as the most significant space activity. It also arises from the above examination that space launch services are not even taken into account as producers of economic benefits.[1015]

6.2.c. Objectives

The National Space Plan enumerates the main objectives of Argentine space policy. An examination of these objectives will also help us assess the existing space policy framework and will contribute to the corroboration or refusal of the premise under analysis, i.e., whether or not Argentine national space legislation is comprehensive and adequate. This analysis may only be done in the context of the purpose for which the Argentine space agency, CONAE, was created.

[1015] For the drafters of the National Space Plan, the social and economic benefits of all space endeavors carried out within the framework of the National Space Program must be wide and immediate. To achieve this result, the National Space Plan proposes to concentrate all efforts on the following: (i) contributing to education and to improving the population's quality of life, particularly in distant and marginal areas; (ii) creating new capabilities and sources of employment at industries producing goods with a high added value; (iii) obtaining advantages for the local production system, thus allowing for the insertion of our industry into a highly competitive international market; and (iv) promoting new businesses that allow for an expansion of the national production scope.

6.2.d. Short and mid term objectives

The principal short-term goals as contained in the original text of the National Space Plan[1016], which covered the 1997-2000 period, aimed mainly at developing and improving the quality of available space data and information. Accordingly, during this time, CONAE assigned most of its resources to remote sensing activities. With respect to space launches referred to in the Plan as access to space, the Plan merely foresaw the implementation of preliminary design studies. However, no actual task regarding space transportation was carried out during the first period of the Plan.

Mid-term objectives, which were conceived to be carried out during the first six years as from the year 2001, tended to improve space means for the obtainment of space data. The most important mid-term objectives dealt with the expansion of ground infrastructure for remote sensing satellites and the commercialization of remote sensing systems[1017]. The most striking feature of mid-term goals was the absence of a program concerned with the development of an expendable launch vehicle. Under the National Space Plan, as originally conceived, the actual building of an expendable launch vehicle was considered only as a long term goal, i.e. a task to be carried out after the year 2006.

[1016] They may be summarized as follows: (a) land infrastructure: (i) construction of compact ground stations, (ii) expansion of existing tracking, telemetry and control ground stations, (iii) operation of multipurpose ground station, (b) Satellite systems: (i) SAC-B Mission, (ii) SAC-C Mission, (iii) preparation of the SAOCOM A/B-1 Mission (Remote Sensing and Telecommunications Satellites), (iv) preparation of the SAC-D Mission (spectrometry), (v) design of the modular satellite engineering concept, (c) data systems: (i) image processing, (ii) interface of geographical information systems, (iii) extension of computer networks, (d) access to space (launch vehicles): (i) design of the conceptual engineering of new generation spacecraft, and (e) institutional development and basic tasks: (i) research and development programs, (ii) cooperative programs with Mercosur countries and (iii) commercialization of SAOCOM Missions.

[1017] They may be summarized as follows: (a) land infrastructure: (i) expansion of infrastructure of Integration and Test Laboratory, (ii) testing of space vehicle subsystems, (iii) ground infrastructure for SAOCOM 1-2 use, (b) satellite systems: (i) SAOCOM 1 A/B-1 Mission, (ii) replacement of the SAC line satellites, (iii) SAOCOM 2 A/B-2 Missions, (iv) replacement and expansion of the SAOCOM series, (c) data systems: (i) advanced geographical data systems with detailed positioning. Interface with on-ground mobile receivers, (d) access to space (launch vehicles): (i) operation and tests of the subsystems for a light space launch vehicle, and (e) institutional development and basic tasks: (i) promotion of space exports, (ii) commercialization of remote sensing systems. National Space Plan at 18

This contrasts CNIE's objectives, which included the construction of an expendable launch vehicle among its top priorities. In effect, the CNIE, which had been created in 1960 within the Argentine Air Force, was actively involved in several projects related to space transportation, such as the EXAMENET sounding rockets[1018] and the controversial Condor II program, a project aimed at developing the first Argentine multipurpose space launch vehicle[1019]. Like most space launch vehicles, the Condor could be used for both military and civil missions. The original idea behind this project was to permit Argentina to be a member of the exclusive club of nations with space launch capabilities[1020]. However, after the Malvinas (Falkland Islands) war[1021]

[1018] In August 1965 the CNIE, together with the *Comissao Nacional de Atividades Espaciais* (CNAE) of Brazil and NASA established a program of meteorological research with sounding rockets named Experimental Inter-American Meteorological Rocket Network (EXAMETNET)˙ Under these programs several launches were conducted from Argentina under the auspices of CNIE. After these fruitful experiences, in the late 1970's CNIE officials started to work on the design of a multipurpose expendable launch vehicle under a strictly confidential basis. In 1981, the Argentine Air Force officially acknowledged the existence of this project, which it called the Satellization Plan, but which was known as the Condor Program. Its objective was to build a multipurpose space launch vehicle. During the first year of this program, there were 15 launchings of HASP and ARCAS rockets in Brazil with full participation of the CNIE. During the second year there were 17 launchings of meteorological sounding rockets from Barreira do Inferno at Natal and two launches took place in Chaco, Argentina, under the auspices of CNIE. Difficulties with procurement precluded the agencies from executing a more frequent schedule of operations. Two Titus rockets were launched successfully on November 12, 1966, from Chaco in Argentina. 12 Nov 1966 #ARG Titus 01 FU150 Astronomy (eclipse) S (274 km) 12 Nov 1966 #ARG Titus 02 FU150 Astronomy (eclipse) S (270 km). The program was divided into three phases. The goal of phase I was to develop the necessary technology for the propulsion of the launching of payloads in low earth orbits (100 and 300 kilometers). The second phase –known as Condor II- consisted of the construction of a two-stage expendable launch vehicle capable of placing payloads of up to 400 kilograms in orbits of around 550 kilometers. The third phase aimed at developing a modular vehicle together with ground support and complete construction facilities to manufacture a series of such vehicles. http://www.met.inpe.br/htmldoc/conferen.html accessed on October 31, 2000.

[1019] "Argentine Space Law and Policy" *supra* note 421 at 177.

[1020] The Argentine Air Force had identified as potential launch bases the Experimental Launch Center of El Chamical, located in the province of La Rioja and the Atlantic Experimental Center located in Mar Chiquita, Buenos Aires. Additionally, Argentina could use the Marambio Base in Antartica, where the CNIE had been sending personnel since 1979. Barcelona & J. Villalonga, Relaciones Carnales, *La verdadera historia de la construcción y destrucción del misil Cóndor II* (Buenos Aires: Planeta, 1992) at 21.

the Condor II project received wide support from Air Force officers who viewed it as central in a military strategy for recouping the Islands.[1022] The United States feared that Argentina might use the Condor II to attempt to recover the Malvinas Islands and, more importantly, that Middle East countries used the Condor II against Israel[1023]. Thus, the United States engineered an international pressure campaign to abort this program[1024]. As a result of this international pressure, the Argentine government halted the Condor II, thus giving up its aspiration of developing its own advanced technology for acquiring space launch capabilities. Thus, the Executive Branch prescribed that all components, parts, and facilities used in building the Condor II were to be "deactivated, dismantled, converted or made unusable."[1025] It further dissolved the CNIE and created CONAE.[1026] This explains the reason why the National Space Plan ignores space transportation

---

[1021] B. M. Mueller, "The Falkland Islands: Will the Real Owner Please Stand Up, (1983) *58 Notre Dame L. Rev.* at 616.

[1022] For these Air Force officials, who actively participated in the program, the Condor II was an ambitious missile and military development plan. Serious budget cuts produced as a consequence of Argentina's defeat in the War, led Air Force officials to look for funds in foreign countries. Soon various countries in the Middle East were interested, including Saudi Arabia, Egypt, Libya, and Iraq. It is believed that both Syria and Egypt advanced or invested money with the view to acquiring the missiles. B. Madani, "New Report Links Syria to 1992 Bombing of Israeli Embassy in Argentina", Middle East Intelligence Bulletin, Vol. 2 No. 3, March 2000, http://www.meib.org/articles/0003_s1.htm accessed on November 1, 2000.

[1023] Argentina and Egypt agreed to cooperate on Condor II in the early 1980s, with Iraq helping Egypt under a secret exchange agreement believed to be known as Badr 2000. All three countries had been helped by a group of European companies known as the Consen Group.

[1024]As part of this strategy, for example, Israel pressed Argentina to end the relationship with Cairo, in return for which Israel would deliver 12 promised A-4Q aircraft. Nuclear and Missile Proliferation (Senate - May 16, 1989), Congressional Records, [Page S5444]; Barcelona & Villalonga, *supra* note 1020 at 21.

[1025] Argentine Decree 995/91.

[1026] The Argentine Executive Branch. CONAE depended directly and exclusively upon the President of Argentina and was transferred the property and scientific personnel of the CNIE. However, after five years of CONAE's creation and when it was certain that the Condor II program was completely dismantled CONAE was further transferred to the Culture and Education Ministry[1026] as such direct control by the President was no longer necessary. CONAE was later transferred to the Ministry of Foreign Affairs. As part of the negotiations between Argentina and the United States for the termination of the Condor II Program, both States executed a space cooperation agreement, which included the launch of the SAC-B, an Argentine scientific satellite.

and why it does not provide any policy basis for the regulation of space launch activities.

6.2.e. Long term objectives

The long term objectives of the National Space Plan are: "the improvement and expansion of means in orbit and the use of space as a working environment: access to space and development of recoverable light satellites"[1027]. As can clearly be seen, the long term objectives are vaguely described and uncategorized. They appear to be simple general guidelines for CONAE's future actions rather than actual goals to be attained. Furthermore, reference to the "improvement and expansion of means in orbit" is not even clear and could be interpreted as encompassing different programs.

The National Space Plan was revised for the first time in 1997[1028] and its scope was extended up to 2008. The most significant change introduced is the express inclusion of Argentina's participation and cooperation with Lockheed Martin in the implementation of the SSTO project[1029]. In 1997 Lockheed Martin and CONAE executed a Memorandum of Understanding whereby Lockheed Martin agreed to subcontract CONAE for the construction of the South American infrastructure for the SSTO project[1030]. Although quite remote, this possibility triggered off a main change in attitude of CONAE towards space transportation, thus giving it more prominence within its programs. This, however, has not been enough to change CONAE's apprehension of space vehicle projects generated outside CONAE's sphere.

[1027] National Space Plan at 19.

[1028] Each revision process entails the evaluation of the actions of the previous two years and it also takes into account Argentine current space capabilities and needs. Each revision begins with a technical auditing followed by the corresponding defense and evaluation by CONAE's Board of Directors. Then the Coordination Committee prepares a revision proposal for the Program. This revision is circulated among CONAE's former Directors and other authorities, reformulated according to their suggestions, and the final version is written by the current Board of Directors to be then submitted to the Executive Branch for its and approval. National Space Plan, First Revision, 1997.

[1029] The 1997 reformulated the so called Access to Space by including the following actions to be carried out: (i) launchings included in the missions foreseen in the National Space Program will be continued, (ii) new generation vehicle. In this case, the actions carried out during the past biennium will be continued with the purpose of developing alternatives for access to space having a high operational probability in the long-term.

[1030] http://www.conae.gov.ar/plan/planespacialc.html accessed on December 14, 2001.

Furthermore, CONAE's possible participation in the SSTO project has been considered as compensation or reward for CONAE's successful control of such outside projects[1031]. Thus, the new language[1032] introduced in the National Space Plan as a result of the 1997 revision makes it clear that the access to space projects should be "in full accordance with current national and international technologies, complying with Argentine foreign and non-proliferation policies, and with international agreements on this subject signed by the Argentine Republic [...] and these developments will have to be carried out within a completely transparent framework, giving priority to associations with national and international entities from countries adhering to the Missile Technology Control Regime (MTCR), mainly the United States and the Federal Republic of Brazil.[1033]"

As can be seen from the above enumeration of objectives, the main areas of space activity scheduled in Argentina involve both scientific-technological research work [1034] and applications. As noted throughout the analysis of the National Space Plan, the main application activity is remote sensing[1035], and then on a considerably lower step telecommunications. CONAE's role with respect to telecommunications is only marginal as it only attempts to serve the telecommunications needs not covered by existing commercial endeavors[1036]. With respect to space transportation projects,

---

[1031] Barcelona & Villalonga, *supra* note 1020 at 22.

[1032] The 1997 Revision also identified the following tasks which were being executed at the moment of the revision: (i) evaluation of conceptual alternatives for new generation vehicles and definition of the configuration of the Launching System; (ii) techno-economical feasibility study for developing a new-generation vehicle; (iii) assembly of the Ground Infrastructure for subsystem level testing and (iv) participation in engineering tasks involved in the project SSTO.

[1033] National Space Plan First Revision, 1997 at 1.

[1034] These are considered to facilitate the preparation and management of space missions, and the design of satellites and onboard instrumentation.

[1035] The most important scheduled activities deal with remote sensing and CONAE's efforts are directed towards developing a remote sensing program. This includes the reception and distribution of images from international systems and the generation of image and data through the launching of the series of SAC (Satellite for Scientific Applications) satellites.

[1036] The National Space Plan also contains twelve annexes dealing with different aspects contemplated in the Plan. Annex XI, entitled Legal and Institutional Topics, sets several objectives and makes recommendations concerning legal questions. In this respect, the Plan stipulates that serious consideration should be given to the analysis of the convenience of acceding to the Moon Agreement, and recalls that this Agreement has arisen from the proposal of the Argentine representative in the United Nations. CONAE has so far not taken the initiative to recommend the Executive Branch to send the Moon Agreement for discussion in the Argentine Congress, thus

except for one specific program, where CONAE is the sole Argentine participant, the National Space Plan restricts and disincentives all other possible projects.

6.3. Executive Branch and agency resolutions

Other legal norms concerning the regulation of outer space activities include the norms dealing with the implementation of the Argentine National Registry of Objects Launched into Outer Space,[1037] the creation of the Space Operators Registry,[1038] the Disclosure of Satellite Activities[1039], and a series of resolutions adopted by the communications authorities aimed at restricting the satellite telecommunications market.[1040] These norms will be extensively scrutinized in the section addressing the implementation of the authorization and supervision principle and the implementation of registration obligations. Thus, here we will only examine these norms very briefly with the sole purpose of assessing whether they provide an adequate general legal framework for the regulation of all space activities.

The norms concerning the National Registry of Objects Launched into Outer Space regulate the creation and operation of this registry. They mandate that all acts related to a space object launched or promoted by the

---

Congress has officially not yet even been provided with a copy of this treaty. The Plan recommends the Congress to ratify the Convention Relating to the Distribution of Programme-Carrying Signals Transmitted by Satellite, May 6-21, 1974, 13 I.L.M. 1444, and to enact the domestic legislation required by the Convention. According to the Director of Legal Affairs of CONAE "this has not done simply because this Convention is not in fashion any more". The Plan in Annex XI also puts forward the necessity to create the Argentine Space Object Registry and the Space Operators Registry, which have already been established. It also proposes the evaluation of the inclusion of the analysis of the Settlement of Space Law Disputes in the United Nations agenda, recalling that this topic was studied for the first time in Argentina in 1979. Again, very little, if any, has been done in this respect. The National Space Plan recommends that CONAE give support to the Executive Branch in international forums devoted to the creation of Space Law and to contribute to the training of Space Law experts. In this respect, CONAE prepares some reports to assist the Argentine delegates to COPUOS, which are permanent staff of the Ministry of Foreign Affairs with no knowledge on Space Law or space science. It is CONAE that prepares the responses to various questionnaires sent by COPUOS. CONAE has also given its position with respect to the discussion and analysis of the treaties related to Outer Space, such as the UNIDROIT proposal.

[1037] CONAE Resolution 252/96 on June 8, 1996 and Argentine Decree 125/95.

[1038] CONAE Resolution 463/97.

[1039] CONAE Resolution 330/96.

[1040] "Satellite Reciprocity", *supra* note 526 at 38.

national State or launched from its territory or facility, whether by the Sate itself or private entities, are to be recorded in the National Registry. The resolution also prescribes the content of the information to be provided for recording and establish that CONAE must operate this registry[1041].

Resolution No. 330/96 entitled Disclosure of Satellite Activities requires all public and private entities which begin or develop activities related to satellites, including the design, construction, launching and operation of satellites, to notify CONAE of their engagement in such activities. Additionally, CONAE is to be notified of all tasks carried out in connection with the use of satellites in the scientific field, and for remote sensing and telecommunications purposes[1042].

CONAE also created the Space Operators Registry with the purpose of recording all the institutions, whether private or public, which have plans or programs related to space activities. This registry is inserted within CONAE's Registry Unit[1043].

Finally, the communications authorities issued a series of resolutions which establish a legal regime aimed at restricting competition, both foreign and local, in the satellite telecommunications market[1044]. Thus, for the purposes of entering the satellite market, CNC's regulations have divided the Argentine market into different segments. First the regulations differentiate between geostationary satellite services and non geostationary ones. The former, in turn, are divided into a mixed system of exclusivity (monopoly) and competition according to different radio frequency bands. Special access conditions have been devised for each of these segments, which are examined below[1045].

7. CONCLUDING REMARKS REGARDING THE COMPLETENESS OF THE LEGAL FRAMEWORK DISCUSSION

As arises from the foregoing analyses, Argentine national space laws do not offer an adequate framework for the regulation of space activities. The National Space Plan is notable for its silence on most important matters. As can be observed from our above analysis, under pompous language and

---

[1041] CONAE Resolution 252/96.
[1042] CONAE Resolution 330/96.
[1043] CONAE Resolution 463/97.
[1044] SC Resolution 14/97.
[1045] "Satellite Reciprocity", *supra* note 526 at 38.

sophisticated methodologies, it merely identifies a vague list of potential benefits of space programs and outlines general objectives without providing for adequate legal or policy mechanisms for the implementation of these objectives. Furthermore, the examination of the policy objectives and the rationale behind the National Space Plan's conception of economic benefits clearly demonstrates that the most important instrument of space policy in Argentina does not even purport to have a comprehensive and far reaching scope, and places a disproportionate emphasis on remote sensing activities in detriment of satellite telecommunications and space transportation.

The other Space Law norms deal with specific concerns of the Executive Branch agencies, mainly the disclosure of satellite and related activities and the restriction to the satellite telecommunications market. They do not contribute to a general legal framework for the development of space activities in Argentina and are clearly inadequate to regulate the space activities of the private sector.

Current Argentine space laws, which have been adopted since the late 1990's, do not fit into the present paradigm of space development. They assign the State a central role in the conduct, management and decision making of space activities and relegate the participation of private sector companies to a secondary role. The existing legal framework further discriminates against certain activities, in particular, space transportation, and orients the political efforts and objectives of the whole country –including its private sector- towards a single space area.

Therefore, it is submitted that the existing domestic space legal scenario is inadequate for the regulation of the space activities of the private industry, and it is oblivious to important areas of the space sector, most notably space transportation. In effect, if, for example, a private space launch firm wanted to provide services from Argentina it would not have a clear legal framework. It would not have a licensing system and clear procedures to apply for authorization. It would not have any indication of how to abide by the environmental constitutional prescriptions and what standards they actually require. The same would be true for the satellite sector and any other area of the space industry. Moreover, the policy guidelines formulated in the National Space Plan would not even serve as the basis for the adoption of these norms.

C. RESPONSIBILITY AND LIABILITY

As arises from our discussion of the international legal framework, the international responsibility and liability regimes impose significant burdens on States, for they are held internationally responsible for the activities of their nationals and the launching states are held internationally liable for damage caused by a space object.

We will now examine whether and to what extent the Argentine State adopted a regime which allows it to protect itself from the consequences of the international responsibility and liability regimes. For this purpose, we will analyze whether Argentine authorities instituted a safety mechanism for the authorization of space activities which may generate liability on the part of the State. Additionally, we will assess whether the existing liability norms contained in the Civil Code allow the State to seek recovery of compensation paid pursuant to the Liability Convention from the actual doer of the wrong. Finally, we will discuss whether the Argentine State has adopted an adequate risk management system to permit the fulfillment of their space policy objectives.

1. SAFETY LAWS

The disclosure and notification system adopted by CONAE[1046] does not contain any safety provision which would allow the governmental authorities to refuse authorization should a projected endeavor be considered to threaten the safety of the population or be likely to cause damage which the national State should then compensate.

The absence of clear safety rules and the consequent lack of actual overview of space activities might generate liability on the part of the State should an activity cause damage both at the international level where the Liability Convention applies and at the national plane where it does not[1047]. In

---

[1046] According to this system all public and private entities which project or develop space related activities must notify CONAE of their engagement in such activities and where CONAE grants authorization only to those activities that conform to the National Space Plan.

[1047] The Liability Convention does not apply to: (i) nationals of the launching State, and (ii) foreign nationals who participated in the operation of that space object. According to Bin Cheng, the first exception is an application of a basic principle of International Law which refrains from dealing with relations between a state and its nationals, and the second one is an application of the principle *Volenti non fit jura.* B. Cheng, "Convention on International Liability for Damage Caused by Space Objects",

this respect, in our analysis of the regimes implemented by other States it was concluded that the best method to minimize the potential liability of States was the adoption of a safety regime whereby the State assessed the potential risks which a projected space activity entails as a condition for the issuance of the authorization. Generally speaking, these regimes require the verification of all major technical aspects related to a space activity. For example, in the United States, during the safety review, the competent authority examines the launch site, the quality procedures, the capacity of the personnel and the launch vehicle equipment in order to ensure that the launch will not endanger the public safety of the United States[1048].

At the national level, in cases which are unrelated to space activities but whose holdings may be considered to apply to accidents caused by space objects, the Argentine Courts have held that in general the State has a non delegable duty to oversee the safety of their inhabitants and that the omissions to comply with this duty and to provide adequate safety measures trigger off the liability of the State[1049]. The doctrine of omission of the duty to oversee the safety of their inhabitants has also been widely accepted by scholars and commentators[1050]. At the international level, the Argentine state's lack of compliance with the authorization and supervision of the activities of their nationals in outer space will also generate international responsibility as the lack of control would at least amount to a contravention of the non interference obligations of States as contemplated in the Outer Space Treaty[1051].

Based on the foregoing discussion, it is abundantly clear that there is a great need for Congress to institute a safety mechanism for the authorization of activities in outer space, or as a minimum, to instruct CONAE to develop those measures. Based on the analysis of the regimes of other States it is submitted in the present study that the best alternative to implement this safety regime is to model it after the safety review mechanism contemplated in the Commercial Space Launch Act of the United States, which would provide the Argentine state with efficient instruments and procedures to

---

in Jasentuliyana, Nandasiri and Lee, Roy S. K. (eds.), *Manual on Space Law* (New York: Oceana, 1979) at 101.

[1048] See *supra* Chapter II A.

[1049] Supreme Court of Justice "Morales, Ma. Beatriz v. Provincia de Buenos Aires", Decision 44984, Sep. 8, 1992.

[1050] G. Bidart Campos, "Responsabilidad del Estado por Ejercicio Irregular del Servicio de Seguridad Policial", 152 (1993) *El Derecho* at 208.

[1051] See *supra* Chapter I A.

ensure that the proposed space activities will not endanger the public safety of the Argentine Republic.

2. RECOVERY OF COMPENSATION

It is now necessary to examine whether the Argentine state can recover compensation paid pursuant to the Liability Convention from the doer of the wrong in the case of an accident caused by a space object of an Argentine non governmental entity.

Liability issues in Argentina are regulated in the Civil Code, which was adopted in 1869. Although it was amended frequently, from a substantive viewpoint its text has remained practically intact, except for certain concepts introduced in 1969[1052]. Therefore, the structure, language and legal techniques are quite insufficient to give answers to complex issues, such as the ones derived from activities in outer space. Nonetheless, the Civil Code governs the general rules and principles of civil liability, and therefore it is imperative to analyze its norms, as well as the existing judicial decisions based on its provisions.

The two basic sources of extra-contractual civil liability are based on personal fault as regulated in article 1109 and on the facts of things, also referred to as objective liability in the legal doctrine and jurisprudence, which is governed by article 1113. The Civil Code also contemplates certain cases of vicarious liability.[1053]

3. FAULT LIABILITY

Article 1109 simply states that: "he who, by action or omission, causes harm to another, with fault or negligence intervening, is obligated to redress the damage caused."[1054] From this over-simplistic statement the whole regime of fault liability has been constructed in Argentine law. Basically, liability arises when it can be proved that there is (i) a breach of a law, (ii) damage, (iii) fault or negligence of the defendant, and (iv) a causal link between the damage and

---

[1052] The major reform was in 1969 which introduced the concept of equity, abrogated all of the provisions discriminating against women and incorporated some reforms to responsibility for acts of others.

[1053] The Federal Congress has also enacted a number of laws outside of the Argentine Civil Code that formulate special liability rules.

[1054] Argentine Civil Code, article 1109. *"Todo el que ejecuta un hecho, que por su culpa o negligencia ocasiona un daño a otro, está obligado a la reparación del perjuicio".*

the action of the defendant.[1055] Absent any of these elements, there is no civil liability and compensation may not be recovered.[1056]

This kind of civil liability centers around the concept of fault, which requires that plaintiff prove that the defendant was negligent. This system clearly favors the doer of the action due to the fact that the burden of proof rests on the victim. This was in perfect consonance with the objectives of the industrial development at the time of the enactment of the Civil Code. The need to liberate the industry of excessive burdens was considered essential for its development.[1057] In the case of a space accident, the likelihood of obtaining compensation in a claim brought under article 1109 would be practically non existent[1058].

Under the civil responsibility doctrine, as delineated above, the only legal grounds for the recovery of compensation paid for damage caused by the actual doer of the wrong is the case of joint liability of co-authors. In this respect, article 1109 states that "whenever on account of joint liability, one of the co-authors of the wrong, paid compensation which exceeded its corresponding part, such co-author can file a reimbursement action[1059]." Therefore, in the case of damage caused, for example by a private Argentine company, where there is no fault of the Argentine state, there are no legal grounds for the Argentine state to recover compensation paid at the international level for damage caused by such company. Furthermore, even if there were legal possibilities for the recovery of compensation, the State should prove negligence of the entity that caused the damage, which on account of the standard required by article 1109, would be quite unlikely.

[1055] A. A. Alterini, Lopez Cabana & O. Ameal, *Curso de Obligaciones*, (Buenos Aires: Abeledo-Perrot, 1986) at 94.

[1056] Civil liability doctrine has been developed by the courts and by authors and commentators of the Civil Code. J. J. Llambias; Codigo Civil anotado: Obligaciones en general (Buenos Aires: Abeledo Perrot, 1989); G. Borda, Manual de Derecho Civil: Obligaciones, (Buenos Aires: Abeledo Perrot, 1998) at 293; A. A. Alterini, Lopez Cabana & O. Ameal, Curso de Obligaciones, (Buenos Aires: Abeledo Perrot, 1986) at 104.

[1057] G. Ponzanelli, *La responsabilità civile* (Bologna: il Mulino, 1992) at 52.

[1058] N. Mateesco Matte, *Aerospace Law From Scientific Exploration to Commercial Utilization* (Toronto: The Carswell Company Ltd., 1977) at 159.

[1059] Argentine Civil Code, article 1109: "Cuando por efecto de la solidaridad derivada del hecho uno de los coautores hubiere indemnizado una parte mayor que la que le corresponde, podrá ejercer la acción de reintegro."

4. OBJECTIVE LIABILITY

Article 1113 prescribes that "the obligation of the person that caused damage extends to the damage caused by his dependents or the things which he employs, or which he has under his care. In the event of damage caused by things, the owner or guardian, to be exempted from responsibility, shall have to prove that he has not been negligent, but if the damage was caused by the risk or vice of the thing, he may only be exempted totally or partially by evidencing the negligence of the victim or a third person whom he is not responsible for. If the thing was used against his express or implied will the owner or guardian shall not be responsible.[1060]"

From this isolated provision incorporated in its present text in the 1969 reform to the Civil Code, the whole objective civil liability doctrine has been constructed. Unlike the regime based on fault, the objective liability model does not require the existence of fault or negligence. Therefore, the plaintiff must compensate damage when there is a causal link between the action of the thing and the damage. The objective liability system in Argentina has been loosely modeled after article 1385 of the French *Code Civil*. Under French law, whose *Code Civil* contemplates several circumstances where things may be considered to trigger liability, the development of this doctrine has been done by the Courts, which extended the nature and number of things which may be susceptible to cause damage[1061]. However, unlike French law, Argentine objective liability is centered around a generic concept of damage caused by the things, and therefore, there is no need for the courts to extend this principle to specific cases of damage caused by things not explicitly mentioned in the Code. Therefore, under Argentine law damage caused by a space object under the circumstances contemplated in article 1113 generates liability without the Courts' having to extend or modify this principle.

The ethical foundation of this regime is the existence of the created risk where persons who profit from an injury-producing activity, usually large

---

[1060] Argentine Civil Code, article 1113: "En los supuestos de daños causados con las cosas, el dueño o guardián, para eximirse de responsabilidad, deberá demostrar que de su parte no hubo culpa; pero si el daño hubiere sido causado por el riesgo o vicio de la cosa, sólo se eximirá total o parcialmente de responsabilidad acreditando la culpa de la víctima o de un tercero por quien no debe responder. Si la cosa hubiese sido usada contra la voluntad expresa o presunta del dueño o guardián, no será responsable."

[1061] V. Kayser, Liability risk management for activities related to the launch of space objects: today's environment and tomorrow's prospects (DCL, Thesis, McGill University, 2000) [unpublished] at 303.

business enterprises, should compensate for all injuries inflicted by that activity[1062]. This liability model foresees the necessary existence of a damage, the non necessary presence of fault and the existence of an interest legally protected[1063].

The Civil Code does not allow a third party that paid the compensation on behalf of the liable person to seek reimbursement of that compensation. Therefore, the national State lacks any legal vehicle to recover the compensation paid under the Liability Convention from the person considered liable pursuant to article 1113 of the Civil Code.

5. VICARIOUS LIABILITY

The Civil Code contains some cases of vicarious liability. For example, parents are vicariously liable in civil matters for the acts of their minor children. Other cases of vicarious liability include liability of captains of vessels for the damages caused by the crew[1064], and employers for the acts of their employees[1065].

The traditional justification for vicarious liability emphasizes the right of the principal to control the doer of the wrong[1066]; arguably, for example, a space company could have exercised control over the employee that is in charge of the operations of de-orbiting a satellite which crashes with another satellite due to the negligence of that employee.[1067] Another justification is the doctrine of the creation of the risk[1068], which inspired the 1969 reform to the Civil Code[1069]. A more modern argument would also reflect a policy of distributing losses widely and more effectively. For example, a space company is more likely to be insured against this type of risk, or is more likely to have had the opportunity to obtain insurance than the employee.[1070]

1062 E. A. Tomlinson, "Tort Liability in France for the Act of Things: A Study of Judicial Lawmaking", 48 *La. L. Rev.* 1988 at 1299.

1063 Ponzanelli, *supra* note 1057 at 70.

1064 Argentine Civil Code, article 1119.

1065 *Ibid.* article 1122.

1066 L. Hargrave, "Developments in the Law, 1984-85", *45 La. L. Rev.* 1984, at 397.

1067 The control theory may well be questionable in terms of the realities of the behavior of the doer of the wrong.

1068 Hargrave, *supra* note 1066 at 397. For example, the parents brought the child into existence, guided its development, and produced the kind of individual who caused the damage; they are, therefore, the ones on whom it is more reasonable to place the loss.

1069 Alterini, Lopez Cabana & Ameal, *supra* note 1055 at 49.

1070 Hargrave, *supra* note 1066 at 397.

In article 1123 the Civil Code foresees the possibility of the principal to recover damages paid by its dependents in the event of damage caused by the negligence or fault of the dependents[1071]. At first reading, this provision would seem to permit the State to recover from the actual doer of the action. However, as has been examined above, there is no rule in the Civil Code attributing vicarious liability to the State for the actions or omissions of its nationals and thus this provision does not apply to the State. It is true that the Liability Convention imposes liability to the launching State in the event of damage caused by a space object[1072], but this attribution of liability may not be considered vicarious in terms of article 1123 of the Civil Code.[1073]

Therefore, in the event that the Argentine state pays compensation for damage caused by a national entity, it will not be able to obtain reimbursement of the compensation paid based on the provisions of article 1123 of the Civil Code.

6. CONCLUDING REMARKS REGARDING THE POSSIBILITY OF STATES TO RECOVER COMPENSATION

As arises from the foregoing discussion, Argentine domestic law lacks clear safety rules for the protection of its population and public health. This, together with a lack of actual overview of space activities might generate liability on the part of the State should an activity cause damage both on the international and national planes. Therefore, it is recommended that the Federal Congress adopt a safety mechanism for the authorization of activities in outer space, or as a minimum, that it instruct CONAE to develop those measures. The best alternative to model this safety regime is the safety review mechanism contemplated in the Commercial Space Launch Act of the United States. Additionally, none of the systems of civil liability consecrated in the Civil Code permit the Argentine state to recover compensation paid on account of the Liability Convention from the actual national entity that caused the damage. Therefore, it is submitted in the present study that the Federal Congress should adopt national legislation complementing the provisions of the Civil Code where it should clearly establish that the Argentine state may recover all or part of the compensation paid at the international level from the actual doer of the wrong. Otherwise, the international liability system implies

[1071] Argentine Civil Code, article 1123.
[1072] See supra Chapter I A.
[1073] "Commercial Space", *supra* note 7 at 53.

the assumption of risks and liability of non governmental entities by the Argentine state.

7. RISK DISTRIBUTION SYSTEM

From the foregoing conclusions, it arises that States need to adopt a risk distribution system to reallocate the risks and liability derived from the activities of their nationals, in particular in the space launch sector. Our findings from the examination of the risk sharing regimes adopted by other States indicate that all States that are actively involved in the provision of space launch services have adopted a system where they retain some of the risks derived from space launch activities and re-allocate other risks to the space launch carriers or even to the carriers' customers. This has permitted those States to fulfill their policy objectives in the space arena, whether their goals were to foster the private sector launch industry, the achievement or maintenance of leadership in space or the promotion of the launch service infrastructure[1074].

Therefore, it is submitted that the Argentine Federal Congress should adopt a risk sharing system to allocate the risks arising from space activities. In light of the conclusions reached in our examination of the existing systems, it is submitted that Argentina should first identify its space policy objectives in consonance with general economic and social policy and in conformity with its international obligations. Once this enterprise is done the Argentine authorities should devise a regime which allocates the space risks and assigns liability pursuant to those objectives. This entails a careful assessment of the possibilities of the Argentine state to assume risks in light of the benefits which it may obtain from the development of a robust space industry. In this respect, it is recommended that Argentina model the structure of its regime after the Australian one, as this system provides great flexibility and an efficient use of risk management instruments. The Australian approach, which outlines the guiding principles of its space risk policy in the Law and remits the particulars to the regulations, would also permit the negotiation and introduction of new risk management aspects as the needs arise without changing the law. The actual content of the risk distribution system should be the consequence of a coherent and comprehensive space policy resulting from the consensus of all concerned governmental departments, the space industry, and the general public.

[1074] See *supra* Chapter II.

D. IMPLEMENTATION OF THE AUTHORIZATION AND SUPERVISION PRINCIPLE

1. INTRODUCTION

Based on our conclusions on the foundations arising from the international scenario for the implementation of the international obligations at the national level, we will now examine the Argentine response to these obligations. For this purpose, we will first analyze the extent of the resolution dictated by CONAE to materialize the authorization and supervision principle without taking into account the fact that, as arises from our preceding discussion, this norm is not in accordance with the constitutional regime in Argentina. For this purpose we will resort to the categories of analysis employed in Chapter I for the examination of the international aspects of this principle. Given the fact that there are not any court decisions regarding the application of this resolution, in order to interpret it we will examine the policy guidelines contained in the National Space Plan with regard to CONAE's role in each space project in Argentina. This will allow us to deconstruct the rationale of the norm under examination and to conclude whether it is in conformity with the international space framework. Consideration will also be given to the work of the few existing authors that dealt with this topic. We will also examine whether and to what extent this resolution and the policy guidelines of the National Space Plan follow the general standards and common denominators found in the principal jurisdictions. Finally, since Argentine authorities adopted additional authorization requirements for the provision of satellite telecommunications services, we will also examine this regime in detail.

2. RESOLUTION 330/96

CONAE's Resolution 330 intends to implement the authorization and continuing supervision principle consecrated in article VI of the Outer Space Treaty. Thus, this resolution expressly recognizes that "the Outer Space Treaty holds States responsible for the activities of their governmental and non governmental entities in outer space and that the Argentine state has undertook to authorize and supervise national activities in accordance with article VI of the Treaty"[1075]. Furthermore, the resolution also expressly recalls that "according to the Liability Convention the Argentine State is absolutely

[1075] CONAE Resolution 330/96 Preamble.

liable for damages caused by the space objects launched from its territory or procured by public or private entities that act under its jurisdiction"[1076].

The resolution created a system of disclosure and notification of the initiation of space activities, which CONAE evaluates in order to grant or reject the authorization mandated by the Outer Space Treaty. CONAE's resolution prescribes that "all public and private entities which, by themselves or on behalf of third parties, project, begin or develop activities related to satellites, including the feasibility studies, design, construction, launching and operation of satellites, must notify CONAE of the engagement in such activities. Additionally, CONAE is to be notified of all tasks carried out in connection with the use of satellites in the scientific field, and for remote sensing and telecommunications purposes[1077]". Upon receiving the information, CONAE has to classify it through its Registry Unit and to include it in the National Space Plan[1078]. According to section 2 of the resolution, CONAE is entitled to require further information regarding the activities disclosed to it. Furthermore, CONAE may demand that these entities make the necessary modifications to adjust their activities to the national and international space law regulations, including the National Space Plan. CONAE may not authorize the use, launch and export of the satellites and other objects constructed, elaborated, assembled and/or imported in Argentina to be used in connection with space activities by entities which do not comply with the norms prescribed in the resolution. In the event of non compliance, these entities are also penalized with losing the right to record their space objects in the Argentine Space Object Registry.

3. FORMAL ASPECTS

As arises from our examination of the implementation of the authorization and continuing supervision principle in other jurisdictions, most States have adopted a licensing regime. The requirements of these systems, as well as the procedure, differ in all jurisdictions. Nonetheless, most of these States share some common features, such as the existence of a specific procedure or, at least, reference to the future creation -or delegation of power for the adoption- of a specific licensing procedure, and the establishment of safety measures to ensure that the proposed space activities will not pose perils to the public and its property. However, CONAE envisaged a system where all entities engaged or planning to be engaged in space activities have to disclose their

[1076] *Ibid.* Preamble.
[1077] *Ibid.* article I.
[1078] All information disclosed to CONAE is considered public, unless expressly marked as confidential by the disclosing entity.

engagement and to notify it to CONAE. Thus, in this respect Argentine legislation is unique and differs substantially from all other States. In our discussion regarding the form of the authorization and supervision principle, it was submitted that States are free to implement the form of this obligation. Thus, from a pure formalistic standpoint, the fact that there is no license system does not imply any violation of the international regime. However, in light of our discussions of the existing national legal systems, it is recommended in the present study that the best way to implement this principle is through the adoption of a straightforward licensing regime based on clearly defined requirements in accordance with international law. Furthermore, this regime should also be structured around the common predominant features of the legislation of the States with ample experience in the domestic regulation of space activities and should build upon the World Trade Organization's core principles of transparency and non-discrimination[1079]. This should include a clear, transparent and neutral specific procedure and technical standards to ensure that the space activities will not endanger the public safety of Argentina.

4. SUBSTANTIVE ASPECTS

We will now analyze whether from a substantive standpoint the regime adopted by CONAE's Resolution 330 is conducive to a smooth development of private space activities or whether it creates a hostile legal environment for the Argentine private sector space industry.

5. BASIS FOR THE AUTHORIZATION

The legal grounds for issuing the authorization under Resolution 330 is compliance with international space law and national regulations, including the National Space Plan. With respect to the former, it was submitted that the legal grounds for granting or rejecting the authorization to embark on national activities in outer space is the adherence –or lack of adherence, respectively-to the provisions of the Outer Space Treaty. However, it was also held that in practice, there is no distinction between requiring national activities to conform to the Outer Space Treaty or to all the International Space Law treaties and conventions, as established in Resolution 330.[1080] As arises from our discussion of the international legal regime, this requires the State's

---

[1079] M. A. Warner, "Third Annual Latin American Competition and Trade Round Table: After Seattle: Is There a Future for Trade and Competition Policy Rule-Making?" (2000) 26 *Brooklyn J. Int'l L.* at 328.

[1080] "Commercial Space", *supra* note 7 at 52.

verification that the proposed space activities meet minimum safety standards so as not to interfere with other States. However, nowhere in Resolution 330 or elsewhere in the legislation is there a safety regime or a set of safety standards which a proposed space endeavor must conform to. Absent these express standards, the legal grounds for the issuance of the authorization are either completely discretionary at best or thoroughly inexistent.

The latter ground for granting authorization, i.e. national regulations, including the National Space Plan, requires further analysis. In our analysis of the international legal framework it was submitted that as a matter of internal law States may adopt other parameters or standards for the authorization of a space activity. It is thus necessary to examine the relevant national provisions which an entity planning to engage in space activities must comply with in order to obtain authorization. As discussed above, the only specific domestic norm is the National Space Plan. Pursuant to the Plan, CONAE's role as a space agency is to "design, carry out, control, negotiate, and manage space projects and enterprises". While a simple literal interpretation of this provision of the National Space Plan may lead to the conclusion that this provision refers only to CONAE's projects, in actuality, it is CONAE's view that this mandate includes all projects carried out in Argentina[1081]. Thus, CONAE understands that all space projects in Argentina should be designed, carried out, controlled, negotiated and managed by CONAE.[1082]

While CONAE holds that it welcomes the participation of private companies it intends to reserve for itself a kind of "project management" role and the possibility of deciding which projects are worth carrying out, even with respect to those which have been born out of private initiatives and which do not seek public funds or any public support. For this purpose, Resolution 330 empowers CONAE to request that all public and private entities engaged or planning to engage in any space activity or project make changes to their space projects to comply, among other norms, with the National Space Plan[1083]. In many cases, this may lead to situations where non governmental entities may be impeded to carry out certain activities in outer space. In effect, as examined above, the National Space Plan favors certain activities, especially remote sensing and marginal telecommunications services, and clearly restricts others, such as space launch services. Therefore, for example, a project which intends to develop a space launch vehicle in

[1081] Anibal H. Mutti, Space Policy, Conference at INDAE, July 18, 2000 [hereinafter "Space Policy Conference"].
[1082] National Space Plan at 15.
[1083] *Ibid.* at 16.

Argentina beyond the parameters expressly set forth in the Plan will not be authorized in Argentina.

Therefore, conditioning the authorization to a proposed activity's compliance with the provisions of the National Space Plan creates a situation which hinders the development of a myriad of space activities, particularly in the space transportation sector. As a result, the whole Argentine private space industry loses incentives to develop new space activities and to invest in the development of new space technology. It is recommended in the present study that this notification and disclosure system be replaced by a licensing system as envisaged by most states which have long been involved in the regulation of space activities, and that the Argentine State extend –to the maximum possible extent- the benefits of the freedom of exploration and use principle to private firms and individuals. This would certainly result in a substantial quantitative and quantitative growth in the private space industry, which will have positive repercussions in the whole Argentine economy.

6. ACTIVITIES REQUIRING AUTHORIZATION AND SUPERVISION

CONAE's resolution 330 requires that all satellite related activities, "including the feasibility studies, design, construction, launching and operation of satellites", be notified to CONAE. This enumeration is imprecise and vague. For example, the construction of the satellites, as well as their launching, requires disclosure. However, the norm is silent with respect to the construction of a space vehicle. CONAE has interpreted that this enumeration is merely an example and all activities related to outer space need disclosure[1084]. The reason why the construction of space vehicles and the provision of launch services have not been specifically included is that CONAE understands that these activities may not be authorized because they have not been contemplated as such in the National Space Plan.

With respect to the extent of the activities that need authorization, it was held that only those activities which are actually carried out in outer space fall within the scope of the authorization and supervision principle. However, it was also submitted that this does not imply that States are precluded from enacting national legislation or devising other mechanisms for the authorization and supervision of activities preceding and following or related to those that take place in outer space. Thus, extending the requirement for authorization to preliminary, related and concluding activities

[1084] "Space Policy Conference", *supra* note 1081.

as derives from Resolution 330 is not in itself a violation of the international regime.

7. SUBJECT OF THE OBLIGATION TO AUTHORIZE AND SUPERVISE

It has been submitted that States have ample freedom to adopt the manner of the authorization and continuing supervision of activities in outer space[1085]. Additionally, it was held that in general the authorization and supervision functions have been assumed by an agency or another subdivision of the Executive Branch[1086] and that satellite telecommunications authorizations are granted by the telecommunications authorities and not by the agency which is in charge of authorizing other space activities. In this respect, the Argentine solution is in consonance with the international regime and the common practice of other States involved in space activities.

8. CONTINUING SUPERVISION

Resolution 330 is silent with respect to the supervision of activities in outer space. It is thus necessary to elucidate how the Argentine state complies with this obligation. It was submitted that under article VI of the Outer Space Treaty, States are free to implement the form of the supervision, provided that it is carried out in a continuing fashion. However, it is necessary to examine from a substantive legal standpoint whether or not the methodology adopted by the Argentine state to carry out its supervision obligations creates a favorable environment for the development of the private space industry. For this purpose, we must resort again to the policy directives contained in the National Space Plan dealing with CONAE's controlling and supervising functions.

According to the National Space Plan, once a space project is approved by CONAE –provided it conforms to the space policy guidelines of the Plan- it is included in the National Space Plan, which means, as discussed above, that CONAE may "design, carry out, control, negotiate, and manage space projects and enterprises". This interpretation has been confirmed by CONAE's Director of Legal Affairs, who clearly put forward that[1087]:

[1085] See *supra* Chapter I.
[1086] See *supra* Chapter I B.
[1087] "Space Policy Conference", *supra* note 1081.

> "For CONAE and CONAE's directors the only legitimate space activities are those carried out under the auspices of CONAE. CONAE does not like and does not approve of those private initiatives which have been developed without CONAE's knowledge and blessing. When we learnt that the Argentine Association of Space Technology had been doing so many projects we decided to issue Resolution 330 to try to control and dissuade many of their activities. Unfortunately, we have not been very successful and we are still trying to design our legal framework so that these endeavors will not occur again. Our main concern is how to control and prevent the existence of space initiatives outside or beyond CONAE."

This concern about controlling all space projects and penalizing those that CONAE does not favorably view derives from the objective for which CONAE has been created[1088]. It is evident that CONAE's legal measures are geared toward preventing the emergence of a new space launch vehicle project.[1089] This is further demonstrated by an examination of the National Space Program objectives and the Space Operators Registry Resolution[1090].

The Plan aspires to "attract and incorporate the participation of the private industry, the government, the academic sector and the Argentine science and technology system"[1091]. As put forward in the Plan, "the objective is that the National Space Program be one of the national projects allowing for the concentration of means and resources and for the confluence of diverse human resources towards its accomplishment.[1092]" Again, from this affirmation, it can clearly be seen that CONAE's intent is to drag all private projects and other non governmental endeavors to its area of influence and to

---

[1088] Argentine Decree 2076/94.

[1089] Or any other space program which may disturb the relationship with Argentina's allies Argentina is a major non-NATO ally of the United States. Argentina had launched a diplomatic campaign to achieve this status and once it obtained it Argentina intended to maintain it, even at the expense of aborting its own space industry.

[1090] CONAE Resolution 463/97.

[1091] National Space Plan at 8.

[1092] *Ibid.* at 9.

amalgamate them under the umbrella of the National Space Plan, rather than recognizing their independent existence and permitting them to develop without any interference, provided, of course, they meet international obligations.

Another legal instrument adopted by CONAE to carry out the supervision obligations is Resolution 463/97, which created a Space Operators Registry[1093] This Resolution permits CONAE to centralize the information of all current and potential entities involved in outer space and to control that at all times they will adjust to the National Space Plan[1094]. The information affords CONAE the possibility to have a closer control of the entities engaged in space activities and to require them corrective measures every time they embark on a project which is not contemplated in the National Space Plan[1095]. According to this resolution, all individuals and entities that carry out or plan to carry out space activities must file with the Head of the Registry a request to record the information, which is instrumented in a form included as an annex to the resolution. Once the request is received, the Head of the Registry may require additional information if it considers appropriate.

From the above discussion, it is abundantly clear that the rationale behind CONAE's supervision of space projects is to control that these projects will always conform to the objectives outlined in the National Space Plan, which, creates a hostile environment for the development of the private space industry. As arises from our discussion of the international legal framework, under International Space Law the object of the supervision, i.e., what States have to supervise, is whether the activities of the non governmental entities adhere to the provisions of the Outer Space Treaty. Although States are free to extend this supervision to verify also compliance with national regulations, in the Argentine case the methodology where CONAE practically becomes a major –and unwanted- controlling partner in every space project clearly creates a legal environment that is hostile to the development of its private space industry, and is not compatible with the guarantees consecrated in the Argentine Constitution, in particular the lawful industry guarantee, which clearly prescribes that all the inhabitants of the Nation are entitled to the right to work and perform any lawful industry[1096]. This regime also differs

---

[1093] This registry also depends on CONAE's Registry Unit.

[1094] CONAE Resolution 463/97, article IV.

[1095] "Recopilación y análisis" *supra* note 247 at 52.

[1096] Argentine Constitution, Article 14. "All the inhabitants of the Nation are entitled to the following rights, in accordance with the laws that regulate their exercise, namely: to work and perform any lawful industry; to navigate and trade; to petition the authorities; to enter, remain in, travel through, and leave the Argentine territory; to

substantially from those of other jurisdictions with national legislation space activities. As arises from our discussion of these jurisdictions, the continuing supervision obligation has generally been implemented mainly through the appointment of observers or delegates at the premises of the non governmental entity that conducts space activities, and through the possibility of suspending or revoking a license already issued in the event of the licensee's non compliance with conditions or obligations or in the event of extraordinary circumstances which may jeopardize public health and safety. No jurisdiction has envisaged a supervision scheme aimed at controlling the correspondence of space activities with the arbitrary purposes of the domestic space program as is the case in Argentina.

9. CONCLUDING REMARKS ON RESOLUTION 330

As arises from the examination of Resolution 330, Argentine authorities have created a system of notification and disclosure of current and projected space activities, instead of the widely used licensing system. The Argentine resolution creates a hostile legal environment for the smooth development of private space activities.

10. OTHER AUTHORIZATIONS REQUIRED IN ARGENTINA

Apart from the authorization established under Resolution 330, Argentine law also requires authorizations for the provision of satellite telecommunications services. National regulations on this matter are a series of scattered[1097], often amended and even contradictory norms[1098]. To determine whether these norms comply with the international framework discussed above, we will

---

publish their ideas through the press without previous censorship; to make use and dispose of their property; to associate for useful purposes; to profess freely their religion; to teach and to learn."

[1097] The Telecommunications Act was enacted in 1972 and although it underwent several amendments in practice its provisions are ineffective. The original text conceived telecommunications as a national prerogative subject to a rigid governmental control and the exploitation of satellite services was reserved to the state. Furthermore, the use of antennae for individual reception of satellite broadcasting was forbidden by the Act and only permitted in 1989 through Argentine Decree 174/89, which adopted the principle of free satellite reception, thus allowing viewers to receive television signals directly in their homes. Other important decrees modifying the statutory provisions of the Telecommunications Act were introduced in the late 1980's and the beginning of this decade, such as Argentine Decree 1842/87.

[1098] "Satellite Reciprocity", *supra* note 526 at 38.

closely examine them in the context of the whole Argentine satellite telecommunications framework.

11. COMPETENT AUTHORITIES

The satellite telecommunications regime has been established by the Secretary of Communications[1099] and has been implemented and overseen by the National Communications Commission (CNC)[1100].

CNC has exclusive authority on the issues assigned to it, which implies an overlapping of functions with CONAE[1101]. As discussed above, at least from their statutory by-laws, CONAE also has competency over

[1099] In 1996 the Secretariat of Communications was created through Argentine Decree 660/96. The Secretariat of Communications assumed all its functions and responsibilities. Its main objectives are to: (i) assist the Executive Branch in the elaboration, proposal, and execution of telecommunications policies, supervising the compliance and proposing the telecommunications regulatory framework, (ii) prepare regulations on telecommunications, (iii) give instructions to the CNC, (iv) act as the Argentine signatory of the Operating Agreement Relating to the International Telecommunications Satellite Organization and INMARSAT. According to its mandate, the Secretariat of Communications acts as a policy maker and plays a central role in the elaboration and formulation of policy directives.

[1100] The Argentine National Telecommunications Commission was established in 1990 by Argentine Decree 1185/90 as a response to the needs arising from the new scenario derived from the privatization process aimed at modernizing Argentina's telecommunications infrastructure, in particular the privatization of the national telephone company. CNC also depended directly upon the Argentine President. In 1996, after the privatization of the telephone company was completed, it was transferred to the Secretariat of Communications. Unlike CONAE, it has several functions which are not directly related to Outer Space issues. CNC's main functions deal with the administrative and technical regulation, control and verification of telecommunications in accordance with applicable law and presidential policy.

[1101] CNC is entrusted with administering the radio-frequency spectrum, except for broadcasting matters, and satellite orbits, regulating the provision of satellite services in Argentina, and authorizing the use and establishment of communications satellite systems. Additionally, CNC has been given authority to issue regulations on telecommunications services and to revise any modification to the assignment of capacity in transponders No. 185, 186, 187 and 188 on Intelsat 15 VA F13, having to verify that there are not antitrust behaviors or a discriminatory treatment to the users of such satellite facility. Pursuant to Argentine Decree 1185/90, CNC has to participate in the drafting and negotiation of all international treaties and conventions dealing with telecommunications and technical cooperation matters, and to render advice to the national entity which represents the Argentine state before an international forum.

satellite issues. In fact, CONAE is the only competent state agency to design, execute, control and administer outer space projects and endeavors. For CONAE's officials, this has been interpreted as having the exclusive competence to issue regulations on satellite services. So, satellite operators in Argentina have to comply with the sets of rules imposed both by the CNC and CONAE[1102]. Political tension between the CNC and CONAE amounted to a confrontation between these two organs, which resulted in the adoption of a burdensome double labyrinth of regulations[1103].

12. GENERAL SATELLITE TELECOMMUNICATIONS FRAMEWORK

Argentina expressly excluded geostationary fixed satellite services from its commitments before the WTO[1104], which are precisely the services provided by the predominant satellite domestic operator[1105]. Thus, none of the WTO principles, including those of market access and non discrimination, are applicable to this kind of service. National regulations established a licensing regime for authorizing these services based on restrictions to both foreign and local operators, especially in the Ku band[1106]. Under this regime, for the purposes of issuing the pertaining licenses, CNC's regulations have divided

[1102] "Argentine Space Law and Policy"*supra* note 421 at 177.

[1103] CONAE Resolution 303/97.

[1104] Argentina has signed and ratified the Basic Telecommunications Agreement negotiated before the World Trade Organization and thus it is bound by its provisions, except for those services specifically included in the list of article II exemptions. G. Oberst, "Satellites and World Trade", (1999) *Regulatory Update* at 18.

[1105] For Argentine communications authorities, the most significant political and legal decision concerning satellite telecommunications in Argentina was the establishment of a domestic satellite system. After several unsuccessful projects which began in 1974 under the promotion and supervision of CNC's predecessors, in 1993 Argentina finally succeeded in having its own national satellite telecommunications system. In 1985, in response to the recommendations of one of such projects, Argentina registered two orbital positions for two satellites before the International Telecommunication Union: NAHUEL 1 and NAHUEL 2 (80° W and 85° W respectively). In 1992, through Argentine Decree 2061 Argentine authorities announced a national and international tender to set the national satellite telecommunications system. The project was formally awarded to the winning joint venture, made up of Deutsche Aerospace (Germany), Aérospatiale (France) and Alenia Spazio (Italy), which was then incorporated in Argentina under the name Nahuelsat S.A. Later, General Electrics acquired around 30% of its capital stock, and it is now, in practice, the controlling shareholder.

[1106] This policy is geared toward implementing a legal framework to protect the predominant satellite operator.

the Argentine market into different segments. First the regulations differentiate between geostationary satellite services and non geostationary ones[1107]. The former, in turn, are divided into a mixed system of exclusivity (monopoly) and competition according to different radio frequency bands. Special access conditions have been devised for each of these segments, which we analyze below[1108].

13. GEOSTATIONARY SATELLITE SERVICES

13.1 Ku-Band

Argentine satellite operators enjoy exclusivity in the Ku band[1109]. Furthermore, there is a legal monopoly for seven years to protect the incumbent Argentine satellite domestic system during which period no authorizations may be granted even for other Argentine satellite operators[1110]. During this seven year period, no other satellite service operator may be authorized to provide services in this band[1111].

However, there are certain few circumstances under which foreign operators may obtain authorization to operate in the Ku Band. In principle, these are: (i) when there is no available Argentine satellite capacity, (ii) when Argentine satellite capacity is offered at abusive prices, or (iii) when there are

---

[1107] "Satellite Reciprocity", *supra* note 526 at 38.

[1108] The protections granted to the predominant operator are not restricted only to banning competition. In effect, the predominant has been afforded a series of prerogatives and benefits which will place it in a unique situation that makes it difficult for the minor players and the potential new entrants to compete.

[1109] SC Resolution 3609/99, Argentine Decree 793/99, article 20.

[1110] This reflects the natural monopoly view of the Argentine authorities with respect to the predominant Argentine satellite services provider.

[1111] The regulations deliberately do not make any reference to the starting point of the seven-year period. Taken into account that the license was given in December 1993 the exclusivity period should have finished by the end of 2000. However, in response to rulings required by potential market operators, CNC authorities have clarified that the seven-year period should be calculated as from December 1996, which is the actual launch of Nahuelsat. In our opinion, this lacks any legal value for under Argentine Administrative Law, absent any specific provision to the contrary, any administrative act entries into force upon its publication in the Official Gazette, which took place in December, 1993. The actual launch of the satellite is irrelevant, especially since services were provided long before the launch of Nahuelsat by virtue of a commercial arrangement with Telesat Canada. This decision again shows a clear indication of the determination of CNC to protect the predominant satellite operator, even at the expense of contradicting well established national law.

technical restrictions which impede the satisfaction of the users' demand by Argentine satellites[1112]. The nature of these exceptions indicates that the actual possibilities for foreign satellite providers are slight to nonexistent. Furthermore, CNC officials indicated in several opportunities that at present these conditions do not exist and that it is reasonable to presume that they will not take place in the near future[1113]. At the end of the seven-year period, save for these exceptions, only Argentine satellites may be authorized to provide services in the Ku Band[1114].

There is a fourth possible source of an exception to the exclusivity regime: the existence of reciprocity agreements between Argentina and a foreign government. In order for a reciprocity agreement to be considered as an exception to the monopoly regime and thus permit foreign operators to obtain a license to provide fixed satellite services via geostationary satellites the agreement must afford Argentine operators the same treatment in the other country as the other State's operators and it must encompass DTH services[1115]. At present, there is only one reciprocity agreement in force between the United States and Argentina[1116]. Mexico and Argentina also signed an agreement in 1997 but it has been practically dormant in the Argentine Congress awaiting for ratification[1117]. Argentine officials claim to be negotiating other reciprocity agreements with Brazil and Spain, but so far no actual progress has been reported on these negotiations, which have been extended for at least three years.

It is worthy of note that a literal reading of the regulations indicates that the three analyzed exceptions need the existence of a reciprocity agreement. In other words, under a literal interpretation, even if, for example, there is no available Argentine satellite capacity a foreign satellite services provider may not enter the Argentine market unless its national administration has signed a reciprocity agreement with Argentina. Conversely, the mere existence of a reciprocity agreement would not entitle a service provider to render satellite services in Argentina unless one of the three exceptional circumstances foreseen in the regulations may be evidenced. The text of this provision of the regulations has been the object of severe criticism by US

[1112] SC Resolution 14/97.
[1113] "Recopilación y análisis", *supra* note 247 at 55.
[1114] "Commercial Space", *supra* note 7 at 294.
[1115] SC Resolution 14/97.
[1116] "Satellite Reciprocity", *supra* note 526 at 38.
[1117] *Ibid.* at 38.

Federal Communications Commission which intended that PANAMSAT[1118] be able to provide DTH services in Argentina.[1119] Thus, as a consequence of this pressure, the CNC interpreted that the existence of a reciprocity agreement may be considered as an exception to the exclusivity regime and consequently as grounds for the obtainment of the pertaining license.[1120] Even if the effect of this decision is somewhat more in consonance with the international framework, it contradicts the text of the resolution which the CNC itself elaborated, for, as examined earlier, it establishes otherwise[1121]. However, by virtue of this decision, now a foreign operator whose administration executed a reciprocity agreement with Argentina may provide services in Argentina as if it were a national operator, even in the bands reserved for Argentine satellite systems.

It is still debatable, however, whether the seven-year exclusivity prohibition for rendering services in the Ku Band applies to foreign-licensed satellites, for SC Resolution 14/97, as amended, precluded the issuance of new authorizations until the expiration of that period. In effect, if the reciprocity agreement has the effect of treating a foreign operator as if it were a national service provider, the prohibition to authorize satellite service operators in the Ku band during the seven-year period should also apply to satellite companies admitted to enter the Argentine market on the grounds of a reciprocity agreement. In our opinion, during the seven-year period foreign-

---

1118 SC Resolution 14/97 contemplated a specific regime for those entities that had been providing services in the Argentine satellite market before the commencement of operations of Nahuelsat. As discussed above, these entities were INTELSAT and PANAMSAT. Under this regime, paradoxically denominated continuity regime, INTELSAT's existing licenses were conditioned to the fact that INTELSAT remain an international intergovernmental organization. It is a well know fact that since the early 1990's INTELSAT has been under an irreversible restructure procedure that will end up with the privatization of the organization. Therefore, the existing license will terminate the moment the privatization is complete. For Panamsat, the effects of this resolution were more immediate. It drastically changed the status of its full operational license to a precarious nature. However, as we discussed above, this situation was soon remedied as a result of political pressure exercised by the US government.

1119 Panamsat had been providing satellite services in Argentina by virtue of a license obtained before the adoption of SC Resolution 14/97. However, this resolution converted its license in a new one with a precarious nature, which could basically be revoked at any time. The new license obtained by Panamsat under the reciprocity agreement permits Panamsat to operate under a license which confers to it the rights it had enjoyed before the implementation of the new regulatory framework.

1120 SC Resolution 1361/98.

1121 "Satellite Reciprocity", *supra* note 526 at 38.

licensed satellites should be allowed to provide services in the Ku Band if they qualified within one of the exceptions commented above. After the seven-year period they should also be authorized as if they were Argentine satellites by virtue of the reciprocity agreement. However, the position of Argentine authorities is not clear and they have systematically refused to make any comments in this respect[1122].

13.2 Other bands

For all other bands, i.e., the less profitable radio spectrum frequencies, CNC designed a somewhat less restrictive system, called competition regime. In the C Band, under this so called competition regime, foreign operators may render satellite services, provided that these services are offered exclusively in this band[1123]. Thus, a foreign service provider with satellites operating in the C and the Ku Bands, which includes the vast majority of operators may not offer satellite services in Argentina unless it restricts its services to the C Band.

In all other bands, which includes the Ka Band, authorizations for foreign operators may be issued only on a temporary and conditional basis when services may not be rendered by Argentine satellite systems[1124]. Once an Argentine satellite system is capable of providing services in these bands the satellite operator must cease to market its services and may only continue to honor the contracts which are in force. After all contracts have terminated or expired, the foreign operator will lose its Argentine license[1125].

13.3. DTH Services

DTH services, irrespective of the band, may also be provided only through Argentine satellites. As analyzed above, the existence of a reciprocity agreement may, however, permit foreign companies to obtain a license to offer these services in Argentina.

13.4 Other authorizations required

Another norm issued by the communications authorities imposes the obligation to obtain an additional license[1126]. In effect, apart from the license

---

[1122] *Ibid.*
[1123] SC Resolution 3609/99, Argentine Decree 793/99, article 22.
[1124] *Ibid.* article 22.
[1125] *Ibid.* article 22.
[1126] SC Resolution 14/97.

to provide satellite facilities in Argentina, satellite operators, both national and foreign, must seek a license to qualify as satellite facility providers[1127]. Under the Reciprocity Agreement signed between Argentina and the United States, the Argentine government has not waived the requirement for the obtainment of this license and thus all US operators must seek the authorization for satellite facility providers[1128].

14. NON GEOSTATIONARY SATELLITES

Satellite services which are provided via non-geostationary, non-fixed satellite services, and which do not fall within the category of PCS services, were liberalized on November 8, 2000[1129]. With respect to the Personal Communications Satellite services market Argentina made an exception to the WTO commitments and these services are not open to competition. In this case, Argentine officials declared that CNC will make decisions in "the light of present and future needs"[1130]. In practice, this implies that the PCS market will present severe obstacles for foreign players not only because Argentina might eventually close this market for competition but also because of the lack of a clear regulatory framework governing these services[1131].

15. CONCLUDING REMARKS REGARDING THE AUTHORIZATION SYSTEM FOR SATELLITE TELECOMMUNICATIONS ACTIVITIES

The foregoing discussion shows that, apart from CONAE's authorization, any entity that intends to provide satellite telecommunications services in Argentina must also seek authorization from the communications authorities. These implemented a system which severely restricts the issuance of new licenses, particularly in the Ku Band, where the national predominant carrier operates. This licensing system thus limits the entrance of new operators to this lucrative sector. Thus, the licensing system devised for the authorization

---

[1127] The term satellite facility provider is a concept used in Argentine law that means an individual or legal entity licensed to operate satellite facilities.

[1128] "Satellite Reciprocity", *supra* note 526 at 38.

[1129] Oberst, *supra* note 1104 at 18.

[1130] 1130 "Satellite Reciprocity", *supra* note 526 at 38.

[1131] In 1998 Argentina launched a tender process, which was open to national and international operators. After a bidding competition whose legitimacy was questioned in the courts and widely denounced in the press, the CNC granted one license. In light of this experience it is unlikely that communication authorities will call a bid for a new license in the near future.

of satellite telecommunications services creates virtually insurmountable obstacles for new firms to access this market.

E. IMPLEMENTATION OF THE REGISTRATION OBLIGATIONS

As a consequence of the international Cold War confrontation[1132], States could not agree on an international framework that imposed the obligation to inform to the United Nations any relevant aspect other than mere technical parameters. It has thus been argued that States should mandate additional relevant information through national measures[1133]. In this respect, Cocca understands that the extensive information requested by Argentine authorities serves to fulfill this purpose.[1134] In order to determine whether the requirements for the registration of Argentine space objects are compatible with the international regime and are reasonably in consonance with the national measures implemented by other States involved in space activities we will first examine the content of the national norms dealing with the registration of Argentine space objects. Then we will contrast their salient features with our conclusions and findings of the international and other domestic legal scenarios.

1. ARGENTINE SPACE OBJECT REGISTRY

Executive Decree No. 125/95 created the National Registry of Objects Launched in to Outer Space[1135]. It prescribes that all acts related to a space object launched or promoted by the national State or launched from its territory or facility, whether by the State itself or private entities, are to be recorded in the National Registry. Registration is compulsory and there is public access to the registry[1136].

The following information should be recorded with respect to each space object:

- agreements with other launching states
- date and territory or location of launch

[1132] Goldman, supra note 456 at 32.

[1133] A. A. Cocca, "Preface", in J. Hermida, *Commercial Space Law: International, National and Contractual Aspects* (Buenos Aires: Ediciones Depalma, 1997) at xiii.

[1134] *Ibid.*

[1135] Argentine Decree No. 125/95.

[1136] Space objects may be recorded when they are under construction.

- basic orbital parameters, including: (a) nodal period, (b) inclination, (c) apogee, (d) perigee
- general function of the space object
- name and address of owners and/or operators
- identity of manufacturers
- name of launching company
- information regarding insurance
- name of the individual responsible of the control over the space object
- location and characteristics of the telemetry, tracking and command station
- frequency of transmission of the space station on board
- weight of the space object
- expected lifetime of the space object
- measures taken towards preventing contamination of Outer Space
- estimated time of disintegration of the space object
- identification mark

Additionally, CONAE's Resolution No. 252/96[1137], which governs the procedure for the registration of space objects also established that the following acts need to be recorded. These are: (i) the creation, conveyance, declaration, modification and extinction of ownership and security rights on space objects, (ii) the seizure, encumbrances and liens over the space object, (iii) all acts that in any manner affect the legal condition of the space objects, and (iv) acts which restrict or impede the sale of the space object to its owner, operator or other titleholder.

For the purpose of the present analysis, the above enumerated data can be classified in different categories according to their purpose. The norms thus demand (i) information which the Argentine State must communicate to the Secretary General of the United Nations, (ii) basic technical information, (iii) information geared toward facilitating the compensation of damages caused by registered objects, (iv) information regarding environmental issues, and (v) ownership and collateral information. The first two categories require the data generally demanded by most of the countries that implemented a national registry[1138]. With respect to the third category, the decree prescribes the disclosure of the identity of the manufacturer, the responsible of the space object and information regarding insurance. In this respect, it must be noted that Argentine legislation does not oblige the procurement of insurance, but in

---

[1137] CONAE Resolution No. 252/96.

[1138] http://www.oosa.unvienna.org/SORegister/regist.htm accessed on December 14, 2001.

the event that insurance is obtained, the insurance policies must be filed before the Registry[1139]. Additionally, entities must identify the marks of the space objects, which was one of the issues that originated controversies in COPUOS. At the international level, States are not obliged to mark space objects but have to inform the Secretary General of the United Nations if they do so. All this information is geared toward facilitating the victims of a space accident access to information for the obtainment of compensation. The requested environmental information reflects the constitutional and international obligations assumed by the Argentine State. However, since neither the constitutional nor the international norms have been formulated in unambiguous and precise terms, Argentine registration norms merely request general and broad information. Finally, the information regarding ownership and collateral measures on the space object has been introduced by CONAE's resolution and has not been delegated to it by the Executive Decree[1140]. Therefore, and in light of our preceding discussions on the lack of CONAE's power to adopt general measures in the absence of a formal and express delegation of authority, these measures may not be considered to be binding. In any case, they are meant to collect and centralize information but they do not aim at creating any right or preference for creditors, owners or operators of space objects[1141].

It has been submitted that according to the Outer Space Treaty and the Registration Convention, States have ample freedom to implement most aspects of their national registries. These include the legal structure, the form, the moment of creation, the information to be recorded, and all other substantive and procedural issues. It was also held that in light of the abuse of discretion principle, States must refrain from imposing their nationals any excessive burden which may, in practice, restrict their access to space. As arises from the present examination, Argentine law requires more information than that referred to in the Registration Convention. This additional information complements the information needed to comply with the communication to the United Nations International Registry. It also mandates certain issues which could not reach consensus because of the opposition of a minority of States, but which most of the international drafters intended to include in the Registration Convention. It further prescribes the recording of information aimed at facilitating the payment of compensation under

---

[1139] "A Launch Provider's Perspective", *supra* note 224 at 1.
[1140] CONAE Resolution No. 252/96.
[1141] S. Beltrán de Oliva & A. H. Mutti, "Los Contratos Espaciales y el Registro Nacional de Objetos Lanzados al Espacio Ultraterrestre de la República Argentina", in M. Folchi (ed.) V *Temas de Aviación Comercial y Derecho Aeronáutico y Espacial* (Buenos Aires: ALADA, 2001) at 129.

international obligations and at complying with environmental commitments. All of the required data are easily available and may not be considered as creating excessive burdens. Furthermore, in order to protect the registering entities, all the documentation submitted to the registration authorities remains confidential, even when the actual recorded data is publicly available[1142]. Nonetheless, it bears noting that the regimes generally adopted in other jurisdictions are simpler and contain straightforward measures for the compliance with the registration obligations derived from the Outer Space Treaty and the Registration Convention. In most of these jurisdictions the information to be recorded simply follows the standards contained in article IV of the Registration Convention.

F. CONCLUSIONS

Although the Argentine Constitution does not contain any article relating to outer space, the regulation of outer space activities may be considered a federal matter on account of the interstate commerce clause and existing judicial interpretations in the telecommunications and air transport fields. Additionally, domestic space laws which are enacted as a response to the obligations contained in the international instruments of space law are the exclusive prerogative of the Federal government.

Since from a constitutional law perspective there is no reason to treat maritime issues and space matters differently, for both are activities with a strong international aspect, the constitutional principles, rights and guarantees which were conceived for maritime navigation and commerce may be extended to the space fields.

The constitutional provisions which have a general scope apply to outer space activities. These include the environmental obligations introduced in the 1994 amendment, as well as all the guarantees and rights that protect all Argentine inhabitants and entities, such as the free circulation of goods of national production, the protection of property, and the rights to work and perform any lawful industry, to navigate and trade, and to enter, remain in, travel through, and leave the Argentine territory, among many others.

Despite the constitutional prescriptions, the role of Congress has been neglected to the approval of the international treaties and conventions. The existing domestic space norms emanate from decrees of the Executive Branch

---

[1142] CONAE Resolution 252/96.

and resolutions of its agencies. However, absent a specific delegation of power for the adoption of a specific measure, CONAE may not issue regulatory resolutions. This is so because CONAE has not been entrusted with regulatory powers, the Federal Congress has exclusive prerogative to implement the obligations arising from an international treaty, and the role assigned to CONAE under the National Space Plan is that of coordination and not regulation. It is thus recommended that the Federal Congress assume a more active and preponderant role in the formulation of national space law, including the domestic implementation of the international treaty obligations.

Current Argentine space laws, which have been adopted since the late 1990's, do not fit into the present paradigm of space development. They assign the State a central role in the conduct, management and decision making of space activities and relegate the participation of private sector companies to a secondary role. Space programs that do not coincide with state policy objectives are rejected and the existing legal framework empowers the Executive Branch agencies to discriminate against certain activities, in particular, space transportation. Therefore, the existing national space legal scenario has created a very hostile legal environment for its private space industry. Furthermore, the National Space Plan, which is notable for its silence on most important matters, places a disproportionate emphasis on remote sensing activities in detriment of satellite telecommunications and space transportation. The other Space Law norms deal with specific concerns of the Executive Branch agencies, mainly the disclosure of satellite and related activities and the restriction to the satellite telecommunications market, and do not contribute to a general legal framework for the development of space activities in Argentina.

With respect to civil liability, none of the systems consecrated in the Civil Code permit the Argentine state to recover compensation paid on account of the Liability Convention from the actual national entity that caused the damage. Additionally, Argentine law lacks clear safety rules, which, together with an absence of actual overview of space activities, might generate liability on the part of the State in the event of damage arising from a space activity. It is thus recommended that the Federal Congress adopt –or instruct CONAE to adopt- a safety mechanism for the authorization of activities in outer space. This could be modeled after the safety review mechanism contemplated in the Commercial Space Launch Act of the United States. It is further recommended that the Federal Congress adopt national legislation complementing the provisions of the Civil Code to permit the Argentine State to recover all or part of the compensation paid at the international level from the actual doer of the wrong and it should further adopt a risk sharing system to allocate the risks arising from space activities.

For this purpose Congress should first identify its space policy objectives in consonance with general economic and social policy and in conformity with international obligations. The formal structure of the Australian risk allocation system provides a valuable model for configuring the structure of the Argentine system, as it provides great flexibility, an efficient use of risk management instruments, and the possibility of negotiating and introducing new risk management aspects as the needs arise. The actual content of the risk distribution system should be the consequence of a coherent and comprehensive space policy resulting from the consensus of all concerned governmental departments, the space industry, and the general public.

CONAE's Resolution 330 created a system of notification and disclosure of current and projected space activities, instead of the widely used licensing system. As arises from the international scenario, States are free to implement the authorization and supervision principle. However, as the main grounds for authorizing or rejecting a proposed space activity is whether such activity is included within the arbitrarily selected objectives of the National Space Plan and the supervision of the Argentine authorities is limited to controlling that these projects conform to those objectives at all times, this system is not in consonance with the solutions adopted in most of the jurisdictions which enacted norms specifically focused on space and it creates a very hostile legal environment for the Argentine private sector industry. It is therefore recommended that the best way to implement the authorization obligations is through the adoption of a straightforward licensing regime based on the common predominant features of the legislation of the States with ample experience in the domestic regulation of space activities, and by building upon the World Trade Organization's core principles of transparency and non-discrimination, which should include a clear, transparent and neutral specific procedure, as well as, technical standards to ensure that the space activities will not endanger the public safety of Argentina.

Apart from obtaining authorization under this Resolution, entities that intend to provide satellite telecommunications services in Argentina must also seek authorization from the communications authorities. These implemented a system which severely restricts the issuance of new licenses, particularly in the Ku Band, where the national predominant carrier operates. The licensing system devised for the authorization of satellite telecommunications services has created severe obstacles for the entrance of new firms to the Argentine satellite telecommunications market.

The norms instituting the National Registry of Objects Launched into Outer Space require more information than that referred to in the Registration Convention. This additional information complements the information needed to comply with the communication to the United Nations and it also mandates the recording of information aimed at facilitating the payment of compensation under international obligations and at complying with environmental commitments. States have ample freedom to implement most aspects of their national registries, which includes the legal structure, the form, the moment of creation, the information to be recorded, and all other substantive and procedural issues. All of the data required by Argentine norms are relatively easily available and may not be considered as creating excessive burdens. Nonetheless, the regimes generally adopted in other jurisdictions are simpler and they only request the recording of the information contained in article IV of the Registration Convention.

FINAL CHAPTER:

CONCLUSIONS AND PROPOSAL

Introduction

The emergence of a private space industry in an increasing number of States calls for an adequate domestic legal framework to regulate its complex and sophisticated commercial endeavors in outer space. Countries with an emergent private sector industry and without a specific domestic legal framework or with pre-commercial space age regulations are unable to provide a regulatory solution to adequately deal with these private space endeavors.

Our search for a basic and general common legislative agenda for those countries without a specific domestic legal scenario has been geared by three sets of pivotal hypothesis dealing respectively with the International Space Law basis and foundations for the implementation of domestic space legislation, the development of domestic space law in the most important spacefaring States and the implementation of national space legislation in a paradigmatic country –Argentina, which was analyzed as a case study.

The purpose of this book has been to propose the fundamental regulatory policy basis for a future domestic legislation governing private space activities in those States where their industry has or aspires to have a preponderant role in the pursuit of space activities and which have not yet crafted their national space regulatory framework. As mentioned earlier in the introductory chapter, the purpose was not to draft a bill or to exhaustively define all the issues to be addressed in a domestic regulatory system. This is so because it is our understanding that legislative reform aimed at developing national space law must result from full participation of all space players and those likely to be affected by the reform. Our conception of a legal reform is not limited to the mere modification of existing legal rules or the adoption of a new text crafted in academic circles. Neither is it the result of a one legislative session. It is a multifold dynamic process, which requires a national effort based on high level of State and private sector participation[1143].

[1143] Macdonald, *supra* note 55 at 831. It also requires the definition of a space policy to be integrated with general economic and social policy. Furthermore, for a legislative reform to be successful it should rest on three basic pillars: (i) adequate rules, (ii) appropriate processes through which those rules are made and enforced and

In order to formulate our proposal for the fundamental policy basis for domestic space legislation we will first recall the main conclusions regarding each set of hypotheses. Then, we will proceed to examine the different legal alternatives to structure the proposed legislative scenario and finally we will enunciate the main features of our recommended model for a common legislative agenda. The formulation of our model will be done in light of these conclusions and in accordance with the tenets of legal reform and participatory theory doctrines.

A. CONCLUSIONS REGARDING THE INTERNATIONAL LEGAL FRAMEWORK

Under general Public International Law a State is responsible for any violation of its international obligations when it results from an action or inaction by the government of the State, its political subdivisions or any organ, agency, official or employee acting within the scope of authority. Therefore, a State is generally not responsible for the acts of individuals or other private entities. International responsibility is based on objective standards and the existence of damage is not a condition for the existence of international responsibility. Therefore, international responsibility acts as a tool to enforce standards of conduct imposed on States rather than as a means to allocate risks. The regime of international responsibility for space activities deviates drastically from the general norms of international state responsibility with respect to attribution rules, for States have been assigned international responsibility for national activities in outer space carried on not only by governmental agencies but also by non-governmental entities. Under the Outer Space Treaty a State bears international responsibility for the activities over which it has the opportunity to exercise legal control, i.e., activities which are within the state's jurisaction, whether territorial, quasi territorial or personal.

International state liability has developed rather autonomously from the doctrine of international responsibility and it is based on the proposition that absence of wrongfulness does not preclude the compensation for damage caused by an act of a State. This doctrine has been incorporated to the Liability Convention of 1972. The interpretation of the interplay between the

---

(iii) well functioning public institutions appropriately staffed with trained individuals. Nolon, *supra* note 48 at 726.

responsibility and liability provisions of the Outer Space Treaty and the Liability Convention indicates that under the Liability Convention, States are liable for damage caused by the space objects of their national private entities. This applies even in the cases of States which have not ratified the Outer Space Treaty, since its responsibility principles are considered customary international law.

The Liability Convention attributes international liability to the launching state, whose definition has given rise to some concerns in the legal literature, especially with regard to the concept of procuring state. It is postulated that the decision as to whether a State falls within the category of procuring state is a question of fact, which should be made on a case by case basis in light of the parameters contained in the definition of launching state. Even though article V neglected to include the procuring state among those which may be jointly liable the ample definition of launching state contained in article I of the Convention, together with the general principles of joint liability, suggests that the concept of procuring state is to be read into article V. A literal interpretation of article V would run contrary to the purpose of the whole Convention.

The Liability Convention permits special arrangements between States to redistribute their financial obligations, such as the liability regime adopted for the International Space Station. These agreements are valid only among these States and are not opposable to non participating States.

The dispute settlement regime consecrated in the Liability Convention is inadequate since it does not allow private companies to resort to the dispute settlement procedure directly but only through the States. It is thus recommended that States adopt a specific norm allowing their national private companies and individuals to obtain remedy in an expeditious manner for loss of property arisen from damage caused by space objects. It is further proposed that this norm be based on the Italian act of January 25, 1983. Like the Italian law, this norm should provide effective mechanisms for private companies to obtain compensation from its own State even when this State has not actually obtained compensation of such damage from the launching state by virtue of its inaction on the international plane. Additionally, the decisions of the Claims Commission are not always binding. In this respect, States should be encouraged, as urged by the International Law Commission, to make a declaration to consider the Claims Commission's decisions binding as contemplated in paragraph 3 of the General Assembly's Resolution 2777 (XXVI).

The Liability Convention has structured a dual objective and subjective system of unlimited state liability, which implies an onerous burden for States as it is considered that liability arising from space activities is the most omni-comprehensive liability regime. Therefore, at the national level States must adopt a regime to protect themselves from the consequences of these activities. This protection should be twofold. First, States should structure safety laws or other measures to minimize the risks derived from the space endeavors of their nationals. Second, States should adopt national legislation or other domestic legal measures establishing that States may recover all or part of the compensation paid at the international level from the actual doer of the wrong, for otherwise, this international liability system also implies the assumption of risks and liability of non governmental entities by the States themselves. Therefore, States need to adopt a risk distribution system to reallocate these risks and liability according to their political objectives in the space arena.

Under the *Corpus Juris Spatialis* States are not obliged to pass domestic space legislation. States must, however, comply with the requirements contained in article VI of the Outer Space Treaty, which prescribes that the appropriate State must authorize and supervise on a continuing fashion the activities of non governmental entities in outer space. States are free to implement the form of such authorization and supervision, which may or may not include the adoption of national law. The authorization and supervision principle should be read in conjunction with the responsibility provisions of the Outer Space Treaty, and, thus, the concept of national activities may not be construed to be at the entire discretion of each State. There are no exemptions from authorization and continuous supervision for any activity which falls under the scope of Article VI. The legal grounds for granting or rejecting the authorization to embark on national activities in Outer Space is the adherence –or lack of adherence, respectively- to the provisions of the Outer Space Treaty. The same applies to supervision. States may, however, adopt other parameters or standards for the authorization of a space activity. There may be several "appropriate states" in each activity that takes place in outer space. The determination of the appropriate state must be done on a case by case basis in light of the specific characteristics of each activity. The recommended approach is to examine each activity and to identify all those States that may be concerned with it. Within the appropriate state, the question of who must grant the authorization is left to the discretion of that State. Whenever a State is substantially involved in a space activity it is not obliged to expressly authorize and supervise that activity. In this case, its participation can be considered as the authorization and continuing supervision prescribed by article VI of the Outer Space Treaty. In light of the purpose of the authorization and supervision principle, authorization may only be granted or refused before the beginning of said activities.

The freedom of exploration and use principle grants States ample rights to use and explore outer space. It is recommended that States whose general space policy includes the fostering of private space activities extend this freedom –to the maximum extent possible- to their private firms by means of national law.

The content of the principle of authorization and continuing supervision is molded by other principles and norms contained in the outer space treaties and conventions, in particularly the freedom of exploration and use principle.

The principles of: (i) benefit of mankind (ii) non appropriation, (iii) peaceful purpose, (iv) avoidance of harmful contamination, and (v) non interference do not impose, in practice, any serious limitation to the vast majority of space activities. The international cooperation doctrine does not impose any obligation at the national level, as it is considered that it merely constitutes a beneficial advantage for space activities. Under general international law, the abuse of rights principle prescribes that States may not exercise a right in order to cause damage to another State or in a way which may impair the rights of other States. There are several manifestations of the abuse of rights, which encompass the fictitious and the malicious exercise of a right.

International law imposes limitations on States, which reduce their ability to adopt national legislation. These restrictions include the abuse of discretion and the interdependence of rights and obligations. The abuse of discretion doctrine obliges States to exercise a discretionary power in good faith, reasonably, in conformity with the spirit of a treaty, and with due regard to the interests of others. The interdependence of rights and obligations precludes a State from contravening an assumed international obligation on account of national law.

The *Corpus Juris Spatialis* has devised a mandatory system for the registration of space objects. States are free to implement the registry by means of several legal mechanisms, which do not necessarily include the enactment of space legislation. States may establish their registries whenever they are in a position to use it on a frequent basis. There are no international constraints with respect to the form and content of the national registry. States are also free to decide what information must be recorded in their national registry. If possible, States should make the notification of the required information to the Secretary General of the United Nations before the beginning of the space activity. The transfer of satellite ownership in orbit is legally possible under international law both to a launching state and to a non launching state. This requires an agreement among all the launching states for the transfer of the jurisdiction and control rights and obligations. In light of customary and conventional international rules on effects of treaties to third parties, in the case of transfers to a non launching state, acceptance in writing of the non launching state is also required.

B. CONCLUSIONS REGARDING THE NATIONAL LEGAL FRAMEWORK

The nature of domestic law varies from State to State. Except for the United States, most countries which play a preponderant role, or which expect to have an active participation, in space activities have adopted a framework law, such as Russia and Australia. These framework laws are generally wide-ranging in nature, they provide a general regulatory scenario without legislating in detail every single aspect of the space industry, they set the policy basis for future regulation and they refer the regulation of specific aspects of space activities to other agencies. Some acts, such as the Russian and the Ukraine laws, lay down the organization of space activities and define the responsibilities of all state agencies involved in the regulation of space activities.

US domestic space law consists of a series of laws and regulations which govern specific aspects of different space activities, as well as several non specific norms which have a direct impact on the space industry. This reflects the US early involvement in the exploration and use of outer space, and its approach to deal with space law issues as its needs arose. US domestic space law is thoroughly comprehensive and covers all possible areas of outer space exploration and use.

The most important space player without domestic legislation specifically focused on space is France. French law applicable to space activities consists of a series of scattered contractual, administrative and regional norms and arrangements which have been adopted for each space program as the needs arose. The French government, which is currently in the process of elaborating specific national space legislation, has considered that its active participation in space activities fulfilled its obligations under international space law. Similarly, Canada lacks a specific and comprehensive regulatory framework to implement its international obligations. Transport

Canada elaborated a proposal for a policy and regulatory framework, which would continue to treat commercial space launch vehicles as aircraft and would confer authority to the Transport Ministry to authorize, oversee and regulate international launch activities under oversimplified and inadequate guidelines.

In those countries with limited participation in space activities, such as South Africa or Sweden, their main legislative instruments do not deal with all significant regulatory aspects. They are generally very succinct and do not even provide the basis for the elaboration of future space policy.

From a law reform and participatory theory perspective, some legislative instruments are the result of extensive negotiations and consultations with local and international interest groups, such as the Australian Act on Space Activities and the US Commercial Space Launch Act. Some States foresee ample participatory mechanisms for the elaboration of regulations, such as the public notice system in the United States, the consultation rounds in Australia and the public consultations in Canada for the development of regulations and regulatory programs. In the United Kingdom the law does not foresee the direct participation of interested parties in the formulation of regulations but these are subject to annulment of either House of Parliament. Besides, the Parliament created a Parliamentary Space Committee to act as a forum of discussion between members of Parliament and the space industry, which has had an active participation in the elaboration of space policy. In other jurisdictions, such as Russia, Ukraine and Sweden, the law gives marginal participation, if any, to interested individuals and entities in the elaboration of norms and regulations for the exploration and use of outer space. The South African law does not contemplate the possibility of procedures allowing those affected by the law to participate in its enactment but it provides many opportunities for the space industry and the general public to be actively involved in the different aspects of the implementation of space policy, in particular through participation in the Council for Space Affairs and its committees.

All major jurisdictions have envisaged a system for the reallocation of state liability arising from the international instruments of Space Law. The structure and the elements of these systems present general common features as they all have been modeled after NASA's. However, their actual content differs substantially among States, as the objectives of these systems are a response to the objectives of the general space policy of each country. The Arianespace system pursues the maintenance of the French (European) leadership in space, the Australian regime intends to attract foreign companies to establish a launch facilities industry in Australia, the British and Russian norms intend to liberate the government of any risk by reassigning liability to the licensee but the Russian legislation also allows the government to assume part of the risks. The United States regime tends to provide its private sector launch industry with a set of norms that permit it to transfer a significantly high degree of risks to its customers and the government.

All States implemented a licensing system for the authorization of space activities of their non governmental entities. The requirements of these systems, as well as the procedure, differ in all jurisdictions. Some States with limited participation in the space field, such as Sweden, merely enunciated the creation of a licensing procedure but did not adopt specific norms to regulate it. In such cases, the legislation either gives ample discretionary powers to the competent authority or refers the actual process and standards to future regulation. In the Russian Federation the implementation of the authorization principle is twofold. First, the government adopted a licensing system for space activities and second it established a system of certification of space technology. Ukraine also has a dual licensing and registration system of space facilities. The legislation of countries which are not highly involved in the exploration and use of outer space, such as South Africa, tends to be drafted in wide and imprecise terms.

All states which developed norms to govern the award of licenses created a set of safety measures to ensure that the proposed space activities will not pose perils to the public and its property. These technical measures, which are considered the cornerstone of all licensing procedures, vary from country to country. In general, they require the verification of all major technical aspects related to the launch. For example, in the United States, the safety review, which is the most important stage of the license procedure, constitutes a mechanism whereby the competent authority examines the launch site, the quality procedures, the capacity of the personnel and the launch vehicle equipment in order to ensure that the launch will not endanger the public safety of the United States.

In some jurisdictions, notably the United States and the United Kingdom, satellite telecommunications are governed by a sophisticated and specific set of rules and are licensed under specific procedures.

The continuing supervision obligation has been implemented mainly through the appointment of observers or delegates at the premises of the non governmental entity that conducts space activities, such as in the domestic laws of the United States, Russia, Australia, and South Africa, among others, and through the possibility of suspending or revoking a license already issued in the event of the licensee's non compliance with conditions or obligations or in the event of extraordinary circumstances which may jeopardize public health and safety as is contemplated in the national legislation of the United States, Russia, South Africa, Sweden, and Australia.

All countries adopted simple and straightforward measures to comply with the registration obligations derived from the Outer Space Treaty and the Registration Convention. In most jurisdictions the information to be recorded follows the standards contained in article IV of the Registration Convention. The United Kingdom adopted a double register system of space objects composed of a main registry and a supplementary register. The former records all space objects except for those whose launch has been procured by a UK satellite supplier, but which appear on the registry of another State. Under Ukrainian law, space facilities, which include not only space objects but also ground infrastructure, must be recorded in the State Register of Space Facilities. Canada follows a very informal procedure where the Foreign Affairs Department transmits the information prepared by the Canadian Space Agency to the Secretary General of the United Nations through the Canadian UN desk.

Recent studies, which advocate for harmonization of the licensing system in the space arena, fail to consider the already existing harmonizing guidelines, faculties and constraints imposed by the international legal framework. Differences between laws of different countries within the margins set by the international legal regime are generally legitimate on the basis that they are justified by differences in the substantive concerns such as, the stage of technological development and availability of technology; national preferences, geographical characteristics, institutional structures, and the ability of different interest groups to organize, formulate and represent policy objectives.

Despite the conclusions of several recent studies recommending harmonization of national space laws in the area of state indemnification, it is not desirable to harmonize the risk sharing systems, for they reflect the political objectives of the States in the space arena. Harmonization would imply unification of space policies and the consequent elimination of national space objectives. Due to the consequences derived from international obligations assumed by States, the structure and main elements of the risk allocation regime tend to converge in all jurisdictions, especially in those actively involved in space activities.

The international treaties and conventions dealing with outer space provide a framework which is not altogether responsive to the needs of the private sector, especially with respect to impossibility of making direct claims for compensation under the Liability Convention or the ample mosaic of activities which require authorization and continuous state supervision. However, this scenario is unlikely to be changed in the next few years because of the lack of political will on the part of the main spafaring nations. Nonetheless, the experience of States which enacted national space legislation shows that new international rules or new interpretations of existing ones are not needed to fill perceived legal vacuums in the international space legal framework, for they risk imposing new and unnecessary burdens to the space industry. National legislation should be encouraged in those States which have not yet formulated a comprehensive legal framework for the regulation of space activities. These States should profit from the ample experience of States with domestic space law. It is recommended that countries without national law base their future legislative instruments on the common predominant denominators of the national laws of the examined countries as recommended throughout our proposal and in light of Law Reform and Participatory theories.

C. CONCLUSIONS REGARDING ARGENTINE SPACE LAW AND POLICY

Although the Argentine Constitution does not contain any article relating to outer space, the regulation of outer space activities may be considered a federal matter on account of the interstate commerce clause and existing judicial interpretations in the telecommunications and air transport fields. Domestic space laws which are enacted as a response to the obligations contained in the international instruments of space law are also the exclusive prerogative of the Federal government. Since from a constitutional law perspective there is no reason to treat maritime issues and space matters differently, for both are activities with a strong international aspect, the constitutional principles, rights and guarantees which were conceived for maritime navigation and commerce may be extended to the space fields. The constitutional provisions with a general scope naturally also apply to outer space activities. These constitutional provisions include the environmental obligations introduced in the 1994 amendment, as well as all the guarantees and rights that protect all Argentine inhabitants and entities, such as the free circulation of goods of national production, the protection of property, and the rights to work and perform any lawful industry, to navigate and trade, and to enter, remain in, travel through, and leave the Argentine territory, among many others.

Despite the constitutional prescriptions, the role of Congress has been neglected to the approval of the international treaties and conventions. The existing domestic space norms emanate from decrees of the Executive Branch and resolutions of its agencies.

Absent a specific delegation of power for the adoption of a specific measure, CONAE –the Argentine space agency- may not issue resolutions because it has not been entrusted with regulatory powers, the Federal Congress has exclusive prerogative to implement the obligations arising from an international treaty, and the role assigned to CONAE under the National Space Plan is that of coordination and not regulation. The Federal Congress should assume a more active and preponderant role in the formulation of national space law, including the domestic implementation of the international treaty obligations.

Current Argentine space laws, which have been adopted since the late 1990's, do not fit into the present paradigm of space development. They assign the State a central role in the conduct, management and decision making of space activities and relegate the participation of private sector companies to a secondary role. Space programs which do not coincide with the State's policy objectives are rejected and the existing legal framework empowers the Executive Branch agencies to discriminate against certain activities, in particular, space transportation.

The existing national space legal scenario is inadequate for the regulation of private industry space activities and it is oblivious to important areas of the space sector, most notably space transportation. The National Space Plan, which is notorious for its silence on most important matters, places a disproportionate emphasis on remote sensing activities in detriment of satellite telecommunications and space transportation. The other Space Law norms deal with specific concerns of the Executive Branch agencies, mainly the disclosure of satellite and related activities and the restriction to the satellite telecommunications market, and do not contribute to a general legal framework for the development of space activities in Argentina.

With respect to civil liability, none of the systems consecrated in the Civil Code permit the Argentine State to recover compensation paid on account of the Liability Convention from the actual national entity that caused the damage. Furthermore, Argentine law lacks clear technical safety rules for the protection of the Argentine population and public health. The Federal Congress should adopt –or instruct CONAE to adopt- a safety mechanism for the authorization of activities in outer space, which could be modeled after the safety review mechanism contemplated in the Commercial Space Launch Act of the United States. The Federal Congress should adopt national legislation complementing the provisions of the Civil Code to permit the Argentine State to recover all or part of the compensation paid at the international level from the actual doer of the wrong and it should further adopt a risk sharing system to allocate the risks arising from space activities. For this purpose Congress should first identify national space policy objectives in consonance with general economic and social policy and in conformity with Argentina's international obligations. The formal structure of the Australian risk allocation system provides a valuable model for configuring the structure of the Argentine system, as it provides great flexibility, an efficient use of risk management instruments, and the possibility of negotiating and introducing new risk management aspects as the needs arise. The actual content of the risk distribution system should be the consequence of a coherent and comprehensive space policy resulting from the consensus of all concerned governmental departments, the space industry, and the general public.

CONAE's Resolution 330 created a system of notification and disclosure of current and projected space activities, instead of the widely used licensing system. The main grounds for authorizing or rejecting a proposed space activity is whether such activity is included within the arbitrarily selected objectives of the National Space Plan and the supervision of the Argentine authorities is limited to controlling that these projects conform to those objectives at all times. This system is not in consonance with the solutions adopted in most of the jurisdictions which enacted norms specifically focused on space and it creates a very hostile legal environment for the Argentine private sector industry. The best way to implement the authorization obligations is through the adoption of a straightforward licensing regime based on the common predominant features of the legislation of the States with a national space legislation, and by building upon the World Trade Organization's core principles of transparency and non-discrimination, which should include a clear, transparent and neutral specific procedure, as well as, technical standards to ensure that the space activities will not endanger the public safety of Argentina.

Apart from obtaining authorization under this Resolution, entities that intend to provide satellite telecommunications services in Argentina must also seek authorization from the communications authorities. The Communications authorities implemented a system which severely restricts the issuance of new licenses, particularly in the Ku Band, where the national predominant carrier operates. The licensing system devised for the authorization of satellite telecommunications services creates severe obstacles which restrict the access of new firms to the Argentine satellite telecommunications market.

The norms instituting the National Registry of Objects Launched into Outer Space require more information than that referred to in the Registration Convention. This additional information complements the information needed to comply with the communication to the United Nations and it also mandates the recording of information aimed at facilitating the payment of compensation under international obligations and at complying with environmental commitments. States have ample freedom to implement most aspects of their national registries, which includes the legal structure, the form, the moment of creation, the information to be recorded, and all other substantive and procedural issues. All of the data required by Argentine norms are easily available and may not be considered as creating excessive burdens. However, most of the regimes adopted in other jurisdictions, generally based on the standards contained in article IV of the Registration Convention, are simpler and more straightforward.

D. RECOMMENDATIONS AND PROPOSAL

1. ALTERNATIVES FOR THE STRUCTURE OF THE PROPOSED LEGISLATIVE AGENDA

Legal reform could take three different avenues. [1144] First, it can attempt a fundamental change in the current system. For example, this would entail doing away with present Executive Branch agencies which have not been very active, such as the Canadian Launch Safety Office, or which have exceeded the foundations imposed by their constitutive instruments, such as the Argentine CONAE, or abrogating existing national space norms, such as those that impose unnecessary burdens, as the Ukrainian dual licensing and registration system of space facilities, or those that are incomplete and inadequate to deal with private sector space activities, as the Japanese law[1145]. This methodology has been largely criticized and is even considered to be one of the causes of political and regulatory stagnation[1146].

The second avenue consists of changing only those norms and procedures which are incompatible with the international legal framework and leaving intact those norms which are in accordance with the international law, even if they are inefficient or if they have proved to be inadequate to regulate and promote the participation of the private sector in the commercial space

[1144] Nolon, *supra* note 48 at 710.

[1145] This approach has been historically followed by Argentina not only in the space arena but also in every major sector.

[1146] F. Luna, *Breve Historia de los Argentinos* (Buenos Aires: Planeta, 1993) at 104.

market. Although this alternative would be valid from a strictly legal viewpoint, it could lead to a retrocession of current space programs and would deter the implementation of private space endeavors.

The third alternative is the adoption of a framework law within the parameters determined under international law. This framework should define the general space policy objectives, the delimitation of the responsibilities and functions of each governmental authority and the procedure for the adoption of specific regulations that take into account the full participation of all interest groups. Furthermore, it should create new, or reinvigorate the existing, institutions capable of continuously resolving constantly changing situations in the space industry and should ensure the adequate staffing of the agencies charged with implementing and overseeing domestic space measures.[1147]

2. RECOMMENDED LAW REFORM AGENDA

It is recommended that those countries without a –or with an inadequate-specific national space regulatory structure to govern their private commercial space endeavors pursue the framework law avenue. This will avoid the great political barriers to radical reform while creating a reliable scenario within which evolutionary change can occur. This will afford all interest groups the opportunity to resolve the fundamental tensions among them and to codify that agreement in law. The negotiations surrounding the development of that framework can occur as it is discussed and designed and they can also occur each time a single statute or project is proposed under the umbrella of the framework law[1148].

Based on the conclusions made throughout this book, the proposed framework law should have a comprehensive scope and comprise the regulation of all space activities, without favoring any kind of activity over the others. This law should clearly identify the country's space policy objectives in consonance with general economic and social policy and in conformity with international obligations. In those countries with a federal structure, such as Canada, Brazil or Argentina to name but a few, it should attribute –or confirm the existing- legislative competence on space activities to the respective political unit, i.e., the federal state or a province. Given the fact that in general the organization of space activities does not expressly derive from the constitution, the framework law should determine the

[1147] A. Seidman & R. B. Seidman, "Drafting Legislation for Development: Lessons from a Chinese Project", 44 *Am. J. Comp. L.* at 40.
[1148] Nolon, *supra* note 48 at 710.

governance structure and organization of space activities, such as is contemplated in the Russian Federation Law on Space Activity and should distribute responsibilities among different state entities. In this respect, the framework law should distribute the faculties between the Federal Congress – and the provincial legislatures where appropriate- and the Executive Branch agencies with respect to the formulation and implementation of policy in the space arena. Furthermore, it should also contemplate the functions, responsibilities and faculties of the space agency, and other agencies related to outer space matters, such as is prescribed in the Russian Federation law. In this respect, the law should expressly give to the respective space agencies ample faculties for the adoption of regulations of a technical nature.

With regard to non technical norms, the framework law should establish the procedure for the adoption of these regulations, which should contemplate the amplest possible participation of all interest groups concerned in their formulation. In this respect, the US public-notice system, which requires giving the public an opportunity to participate in rulemaking proceedings that might affect private business interests or the personal liberties of private citizens could serve as a point of departure and model. Additionally, the framework law could encourage the incorporation of industry representatives in congressional committees, such as the United Kingdom Parliamentary Space Committee.

In order to ensure high standards in the management of the Executive Branch agencies, the framework law should adopt appropriate and efficient mechanisms to staff the agencies with qualified individuals. For this purpose, it should afford the space industry and the general public ample opportunities to actively participate in the agencies and to be involved in the different aspects of the implementation of space policy. This could be modeled after the mechanism devised in the South African Space Affairs Act.

The framework law must also implement a risk distribution system to reallocate the risks and liability derived from the activities of their nationals, in particular in the space launch sector, which should permit the State to recover from the actual doer of the wrong all or part of the compensation paid at the international level pursuant to the Liability Convention. It is recommended that the structure of the regime be modeled after that of the Australian system –which synthesizes the common denominators of the structure of most risk sharing systems- as it provides great flexibility, an efficient use of risk management instruments, and the possibility of negotiating new risk management aspects as the needs emerge. However, the actual content of the risk distribution system should be the consequence of a coherent and comprehensive space policy resulting from the consensus of all concerned governmental departments, the space industry, and the general public. In order to avoid the problems arisen from the fact that the dispute settlement regime consecrated in the Liability Convention does not contemplate the possibility for private companies to make claims directly to the launching state the framework law should contemplate specific norms to relieve the private companies of the difficulties arisen from the international scenario. It is recommended that these remedies be modeled on the Italian law of January 25, 1983.

The framework law should implement a straightforward licensing regime based on clearly defined requirements in accordance with international law. The system should establish a transparent, efficient and neutral procedure for the request of a license. The license should be awarded to those activities which comply with the requirements arising from international law and which do not impose any safety danger to the State, its inhabitants and their property. For this purpose, it should create a mechanism with clearly defined safety standards, or it should instruct the space agency to develop those standards based on general policy objectives determined in the law. These standards should address the technical conditions and requirements for the launch site, the launch vehicle, and the payload, as well as, the personnel involved in their construction, maintenance and operation. The safety review mechanism contemplated in the Commercial Space Launch Act of the United States provides a useful model to delineate these systems.

The framework law should also regulate the licenses for the provision of satellite telecommunications services based on international telecommunications parameters, such as those arising from the International Telecommunication Union and the common core principles of transparency and non discrimination of the World Trade Organization. This regime should afford the possibility of obtaining a license to all those entities which are technically, financially and legally capable of providing these services. It should eliminate all the prerogatives, if any, that may protect the incumbent satellite telecommunications services provider.

The law should also establish a clear and reasonable continuing supervision regime for all non governmental entities to verify their compliance with international and safety standards. In accordance with the common practice of most of the jurisdictions which implemented domestic space laws, this should be implemented through the appointment of competent delegates with faculties to observe the development of space activities in full respect for rights, freedoms and guarantees of the authorized entity. The supervision should be complemented by vesting the competent authority with powers to suspend and revoke a license in the event of non compliance with conditions or obligations or in the event of extraordinary circumstances which may jeopardize public health and safety.

The framework law should also include the creation of a –or in the case of those countries with an existing registry, the incorporation of the norms dealing with the- national space object registry and should establish a transparent procedure for the recording of all space objects. The information to be recorded should follow the standards contained in article IV of the Registration Convention. Additional information could be required, especially to fill the gaps of the Registration Convention, only to the extent that this does not impose an excessive burden to the licensee.

ABOUT THE AUTHOR

Julian Hermida is a graduate of McGill University's Faculty of Law, from which he holds a Doctor of Laws and a Master of Laws degree. Currently, he is Assistant Professor at Dalhousie University. He has been a Postdoctoral Fellow at the University of Ottawa (Faculty of Law, Common Law Section) and a Visiting Fellow at the University of Cambridge, Faculty of Law, Lauterpacht Research Centre for International Law.

Julian Hermida has over 10 years of experience in negotiating, drafting and managing sophisticated international aerospace agreements, as well as in corporate finance, mergers and acquisitions and other transactional matters focused on the aerospace industry, together with significant regulatory and risk management experience.

Julian Hermida has published and lectured extensively in the United States, Canada, Europe, and Latin America.

Bibliography

- Anthony, Robert A., "Interpretive Rules, Legislative Rules and Spurious Rules: Lifting the Smog, 8 *Admin. L.J. Am. U.*
- Asociación Aeronáutica Argentina, *Informe sobre Política Aérea en Materia de Jurisdicción Nacional y Provincial* (AAA: Buenos Aires, 1985).
- Bahtt, S., Studies in Aerospace Law. From Competition to Cooperation (New Dehli: Sterling Publishers, 1974).
- Balkin, R., "International Law and Australian Federalism" (1998) 92 *A.J.I.L.*
- Ballarino, T. & S. Busti, *Diritto Aeronautico e Spaziale* (Milano: Giuffrè, Milano, 1988).
- Barcelona, Eduardo & Julio Villalonga, Relaciones Carnales, La verdadera historia de la construcción y destrucción del misil Cóndor II (Planeta: Buenos Aires, 1992).
- Barstow Magraw, D., "Transboundary Harm: The International Law Commission's Study of "International Liability" (1986) 80 *A.J.I.L.*
- Barton, "Summary of Discussions", (1967) 10 *IISL.*
- Bederman, D.J., "Revivalist Canons and Treaty Interpretation", (1994) 41 *UCLA L. Rev.*
- Beltrán de Oliva, S. & A. H. Mutti, "Los Contratos Espaciales y el Registro Nacional de Objetos Lanzados al Espacio Ultraterrestre de la República Argentina", in M. Folchi (ed.) Temas de Aviación Comercial y Derecho Aeronáutico y Espacial V (Buenos Aires: ALADA, 2001).
- Bender, R. *Space Transport Liability: National and International Aspects* (The Hague: Martinus Nijhoff, 1995).
- Berwick, T.A. "Responsibility and Liability for Environmental Damage: A Roadmap for International Environmental Regimes" (1998) 10 *Geo. Int'l Envtl. L. Rev.*
- Böckstiegel, K. H., "Developing a System of Dispute Settlement Regarding Space Activities", (1992) 35 *IISL.*
- Böckstiegel, K. H., "The Law Applicable to Contracts" (1982) 25 *IISL.*
- Böckstiegel, K. H., "The Term Launching State in International Law" (1994) 37 *IISL.*
- Bockstiegel, K.-H., "The Term 'Appropriate State' in International Space Law", (1994) 37 *IISL.*

- Borda, G., Manual de Derecho Civil: Obligaciones, (Buenos Aires: Abeledo Perrot, 1998).
- Bourély, "Quelques réflexions au sujet des législations spatiales nationales" (1991) XVI *Ann. Air & Sp. L.*
- Bourély, M., "La Production du Lanceur Ariane", (1981) VI *Ann. Air & Sp. L.*
- Bourély, M., "Rules of International Law governing the Commercialization of Space Activities"(1985) 29 *IISL.*
- Bourély, M., "Rules of International Law Governing the Commercialization of Space Activities" (1986) 30 *IISL.*
- Bourély, M., "Space Law and the European Space Agency" in N. Jasentuliyana (ed.), *Space Law, Development and Scope* (New York: Praeger, 1992).
- Brownlie, I. *Principles of Public International Law,* 5th ed. (Oxford: Oxford University Press, 1999).
- Buergenthal, T., "Modern Constitutions and Human Rights Treaties" (1997) 36 *Colum. J. Transnat' l L.*
- Buergenthal, T., "Modern Constitutions and Human Rights Treaties" (1997) 36 *Colum. J. Transnat'l L.*
- Campbell, N.D., "Australian Treaty Practice and Procedure" in K. W. Ryan, ed., *International Law in Australia* (Sydney: The Law Book Company, 1984).
- Cardozo, B. "Mr. Justice Holmes", (1931) 44 *HARV. L. REV.* at 682.
- Carlino, F., Tesis de Maestría, (Buenos Aires: Flacso, 1997).
- Cassidy, D. E., "Allocation of Liabilities Between Government and Private Sector and Implications on Insurance for Space Commercialization", (1990) 33 *IISL.*
- Castel, J. G., Legal Services provided by the Department of External Affairs with respect to International Judicial Co-operation and Other Matters, in H. M. Kindred, *International Law Chiefly as Interpreted and Applied in Canada*, 6th edition (Toronto: Edmond Montgomery, 2000).
- Cheng, B. "The Commercial Development of Space: The Need for New Treaties" (1991) J. Sp. L.
- Cheng, B., Article VI of the 1967 Space Treaty Revisited: "International Responsibility", "National Activities", and "The Appropriate State", (1998) 26 J.Sp.L.
- Cheng, B., Article VI of the 1967 Space Treaty Revisited: "International Responsibility", "National Activities", and "The Appropriate State", (1998) 26 J.Sp.L.

- Cheng, B., *General Principles of Law as Applied by International Courts and Tribunals* (Cambridge: Grotius Publications, 1987).
- Cheng, B., *Studies in International Space Law* (Oxford: Oxford University Press, 1997).
- Christensen, C. B., "Economics and Regulation of Space Activities" in J. S. Greenberg & H. R. Hertzfeld, Henry R (eds.), *Space Economics* (Washington, DC: AIAA, 1992).
- Christol, C. Q., *The Modern International Law of Outer Space* (New York: Pergamon Press, 1982).
- Clerc, P., "Current French Plans for a National Legal Framework for Space Activities", (2001) 44 *IISL.*
- Clerc, P., "French Policy and Framework" in Project 2001, *Legal Framework for Privatising Space Activities* (Cologne: Institute of Air and Space Law of the University of Cologne, 1999).
- Clift, J., UNCITRAL and the Goal of Harmonization of Law, Internet Law and Policy Forum, Montreal, Canada, 26-27 July 1999.
- Cocca, A. A., "A Way to Complement, Enforce and Improve the Space Treaty and Related International Instruments of Space Law", (1992) 35 *IISL.*
- Cocca, A. A., "Common Heritage of Mankind: A Basic Principle of the International Legal System", (1988) 31 *IISL.*
- Cocca, A A., *Desarrollo progresivo del Derecho Internacional* (Buenos Aires: Consejo de Estudios Internacionales Avanzados, 1991).
- Cocca, A. A., "From Full Compensation to Total Responsibility", (1983) 26 *IISL.*
- Cocca, A.A., "El espacio ultraterrestre. Labor de las Naciones Unidas en la codificación del espacio", in Problemas contemporáneos de la actividad aeronáutica y del espacio, Infante, M.T. and Irigoin, Jeannette, (eds.), *Colección de Estudios Internacionales*, Editorial Universitaria, Santiago, 1978.
- Cocca, A. A., *Mantenimiento de la Utilización del Espacio Ultraterrestre con Fines Pacíficos* (Córdoba: Consejo de Estudios Internacionales Avanzados, 1986).
- Cocca, A.A., "Principles for a Declaration with Reference to the Legal Nature of the Moon", (1957) 1 *IISL.*
- Cocca, A.A., "Recent Developments in Space Law and Space Policy in Latin America", (1996) 16 *ECSL News.*
- Cocca, A.A., "Registration of Space Objects", in N. Jasentuliyana & R.S.K. Lee eds., *Manual on Space Law* (New York: Oceana, 1979) Vol. 1.

- Cocca, A.A., Legal Framework for Economic Activity in Space (Córdoba, Advanced International Studies, 1982).
- Cohen, L. & L. Manion, Research Methods in Education, (London: Routledge, 1994).
- Commercial Space Launch Act Amendment of 1988, Report of the Senate Committee on Commerce, Science and Transportation on H.R. 4399, SR 100-593, Oct. 7, 1988, US Government Printing Office, Washington, 1988.
- Committee on Aeronautical and Space Sciences United States Senate, Convention on International Liability for Damage Caused by Space Objects. Analysis and Background Data, 92d Congress 2d. Session, US. Government Printing Office, Washington, 1972.
- Committee on Foreign Relations, United States Senate, Treaty on Outer Space, 90th Congress, First Session, March 7, 13 and April 12, 1967, (Washington: U.S. Government Printing Office, 1967).
- Couston, M. & L. Pilandon, *L'Europe Puissance Spatiale* (Bruxelles: Bruylant 1991).
- Couston, M., *Droit Spatial Economique* (Paris: SIDES, 1994).
- Cravero, L., *Recopilación y análisis de la normative que regula las actividades espaciales en la Argentina* (Buenos Aires: INDAE, 2000).
- D'Angelo, George V., *Aerospace Business Law* (Westport, Conn.: Quorum Books, 1994).
- Danilenko, G. M., "Implementation of International Law in Russia and Other CIS States," 1998 [unpublished].
- Danilenko, G.M., "The New Russian Constitution and International Law" (1994) 88 *A.J.I.L.*
- Dann, P., "The Future Role of Municipal Law in Regulating Space Related Activities", in T. L. Zwaan, ed., Space Law: Views of the Future (Deventer: Kluwer Law and Taxation Publishers, 1988).
- Davis, M. & R. J. Lee, "Financial Responsibility and Government Indemnities for Commercial Space Launch Activities - The Australian Approach", (1999) 50 *IISL.*
- De Cruz, P., *A Modern Approach to Comparative Law*, (Deventer: Kluwer 1993).
- De Cruz, P., *Comparative Law in a Changing World*, (London: Cavendish Publishing Limited, 1995).
- Dembling, P., "Principles governing the activities of States in the exploration and use of outer space, including the Moon and other celestial bodies", in N. Jasentuliyana & R.S.K. Lee eds., *Manual on Space Law* (New York: Oceana, 1979) Vol. 1.

- Diamond, E., *The Rise and Fall of the Space Age* (New York: Doubleday & Company, Inc., 1964).
- Eisemann, P. M., & C. Kessedjian, "National Treaty Law and Practice: Francen" in M. Leigh & M. R. Blakeslee, eds., *National Treaty Law and Practice* (Washington, DC: ASIL, 1995).
- Ferrer (h), M. A., *Espacio aéreo y espacio superior* (Córdoba: Dirección General de Publicaciones, 1971).
- Ferrer(h)., M.A., "El derecho a la trayectoria", (1997) 13 *IISL.*
- Ferrer, M.A., *Derecho Espacial* (Buenos Aires: Plus Ultra, 1979).
- Filiato, A. R., "The Commercial Space Launch Act: America's Response to the Moon Treaty?" 778 *Fordham International Law Journal*, vol. 10.
- Foley, P., "Trends in Spectrum Management - with Specific Emphasis on Spectrum Pricing Considerations", *QDOS Telecommunications Law*, London, 9-10 February, 1998.
- Forkosch, M.D., *Outer Space and Legal Liability* (Dordrecht: Martinus Nijhoff, 1982).
- Foster, W. F., "The Convention on International Liability for Damage Caused by Space Objects" (1972) The Canadian Yearbook of International Law.
- Frankle, E. A. & E. J. Steptoe, "Legal Considerations Affecting Commercial Space Launches From International Territory", (1999) 50 IISL.
- Galloway, E., "Guidelines for the Review and Formulation of Outer Space Treaties", (1998) 41 *IISL.*
- Gaubatz, W. A., "International Certification for Commercial Reusable Space Transportation", (1999) 42 *IISL.*
- Gerhard, M. & K.U. Schrogl, Report of the Working Group on National Space Legislation, Project 2001, May 2001.
- Gerhard, M. and K.U. Schrogl, Project 2001, Legal Framework for the Commercial Use of Outer Space, Working Group on National Space Legislation, Draft Report at 10 [unpublished].
- Gerhard, M., "Project 2001 Legal Framework for the Commercial Use of Outer Space," Workshop on National Space Law Legislation, Munich 5-6 December, 2000.
- Gillies, D. & R. Marshall, *Telecommunications Law* (Butterworths: London, 1997).
- Goldman, N. C., *American Space Law: International and Domestic* (Iowa: Iowa State University Press, 1988).
- Gorove, S., "Interpreting Article II of the Outer Space Treaty", (1968-1969) 37 *Fordham Law Review.*

- Gorove, S., "Liability in Space Law: An Overview" (1983) VIII *Ann. Air & Sp. L.*
- Gorove, Stephen, "Convention on Registration of Objects Launched into Outer Space. Analysis and Commentary", (1976) 19 *IISL.*
- Greenwood, E., *Methodology of social research* (Buenos Aires: Paidos, 1973).
- Hackworth, Digest of International Law (1943) quoted by H. M. Kindred, *International Law Chiefly as Interpreted and Applied in Canada*, 6th edition (Toronto: Edmond Montgomery, 2000).
- Haeck, L., "Le droit de la guerre spatiale" (1991) XVI *Ann. Air & Sp. L.*
- Hannum, H.,"The Status and Future of the Customary International Law of Human Rights: The Status of the Universal Declaration of Human Rights in National and International Law" (1995/1996) 25 *Ga. J. Int'l & Comp. L.*
- Hargrave, L., "Developments in the Law, 1984-85", *45 La. L. Rev.* 1984.
- Hazelrigg, G. A., "Cost Estimating for Technology Programs" in Greenberg, J. S & H. R. Hertzfeld, (eds.) *Space Economics* (Washington, DC: AIAA, 1992).
- Herczeg, I., "Problems of Interpretation of the Space Treaty of 27 January 1967", (1967) 10 *IISL.*
- Hermida, J., "Argentine Space Law and Policy" (1996) XXI-II *Ann. Air & Sp. L.*
- Hermida, J., "Risk Management in Arianespace space launch agreements", Annals of Air and Space Law", (2000) XXV *Ann. Air & Sp. L.*
- Hermida, J., "Risk Management in Commercial Launches", (1997) *Sp. Pol.*
- Hermida, J., "Satellite Reciprocity Agreement between the United States and Argentina", (1998) 27 *J.Sp.L.*
- Hermida, J., "Space Insurance: A Launch Provider's Perspective" (1997) 11 *The Air and Space Lawyer.*
- Hermida, J., "Space Registry" (1996) 24 *International Business Lawyer.*
- Hermida, J., *Commercial Space Law: International, National and Contractual Aspects* (Buenos Aires: Ediciones Depalma, 1997).
- Hermida, J., *Legal Aspects of Space Risk Management. The allocation of risks and assignment of liability in commercial launch services*, (LL.M., Thesis, McGill University, 2000) [unpublished].

- Hermida, J., *Norms governing launch services by NASA and commercial US private companies*, (LL.D. Thesis, Catholic University of Cordoba, Doctorate of Laws Thesis 2000) [unpublished].
- Hermida, J., Space Financing, (1998) 13 *The Air and Space Lawyer.*
- Hobbe, Stephan, "Global Challenges to Statehood: The Increasingly Important Role of Nongovernmental Organizations" (1997) 5 *Ind. J. Global Leg. Stud.*
- Hosenball, S.N., "The Law Applicable to the Use of Space for Commercial Activities", (1983) 26 *IISL.*
- Hoskova, M., "Intersputnik - New Legal Developments", (1995) 38 *IISL.*
- International Law Commission, Draft articles on Responsibility of States for Internationally Wrongful Acts, 53rd session, 2001, article 4.
- Jackson, J. H., "Status of Treaties in Domestic Legal Systems: A Policy Analysis" (1992) 86 *A.J.I.L.*
- Jackson, J. H., "United States" " in F. G. Jacobs & S. Roberts, eds., *The Effects of Treaties in Domestic Law* (London: Sweet & Maxwell, 1987).
- Jakhu, R.S. & J. Wilson, "The New US Export Control Regime and its Impact on the Communications Satellite Industry", (2000) XXV *Ann. Air & Sp. L.*
- Jakhu, R.S., *The Legal Regime of the Geostationary Orbit*, (DCL, Thesis, McGill University, 1983) [unpublished],
- Jasentuliyana, N., "Opening Remarks", (1995) 38 *IISL.*
- Jenks, C.W., Space Law (London: Stevens, 1965).
- Johnson-Freese, "Alice in Licenseland: US satellite export controls since 1990", (2000) 16 *Sp. Pol.*
- K. L. Li, *World wide space law bibliography* (Toronto: Carswell Co., 1978).
- Kahn, P. ed., *L'Exploitation commerciale de l'espace: droit positif, droit prospectif*, (Dijon: Litec Credimi, 1992).
- Kamenetskaia, Elena, "Cosmos. Cooperación. Derecho" *Ciencias Sociales*, (1983) 1 *Academia de Ciencias de la URSS.*
- Kayser, V., "Commercial Exploitation of Space: Developing Domestic Regulation", (1992) XVII Ann. Air & Sp. L.
- Kayser, V., *Legal Aspects of Private Launch Services in the United States*, (LL.M., Thesis, McGill University, 1991) [unpublished].
- Kayser, V., Liability risk management for activities related to the launch of space objects: today's environment and tomorrow's prospects (DCL, Thesis, McGill University, 2000) [unpublished].

- Kennedy, C.H., & M. Veronica Pastor, An Introduction to International Telecommunications Law, (Boston: Artech House, 1996).
- Kerrest, "Remarks on the Notion of Launching State" (1999) 41 *IISL.*
- Kerrest, "Remarks on the Responsibility and Liability for Damage Caused by Private Activity in Outer Space" (1998) 40 *IISL.*
- Kindred H. M., *International Law Chiefly as Interpreted and Applied in Canada*, 6th edition (Toronto: Edmond Montgomery, 2000).
- Knop, K., "Here and There: International Law in Domestic Courts" (2000) 32 *N.Y.U. J. Int'l L. & Pol.*
- Kolossov, Y., "Outer Space Activities as an Object of International Space" (1985) 28 *IISL.*
- Kopal, Vladimír, "The Doctrine of Space Law", in N. Jasentuliyana ed., *Space Law, Development and Scope* (New York: Praeger, 1992).
- Koven Levit, J., "The Constitutionalization of Human Rights in Argentina: Problem or Promise?" (1999) 37 *Colum. J. Transnat'l L.*
- Krawec, R., "Ukrainian Space Policy Contributed to Natural Economic Development", (1995) 11 *Sp. Pol.*
- Lachs, M., *The Law of Outer Space: An Experience in Contemporary Law-Making* (Sijhoff: Leiden, 1972).
- Laidet, L., "The French Space Program", in W. C. Thompson & S. W. Guerrier, (eds.) *Space: National Programs and International Cooperation* (Boulder: Westview Press).
- Lang, Peter, "La responsabilité internationale encourue en raison des activités liées a l'utilisation de l'energie nucleaire by Angelo Miatello. BOOK REVIEW", (1989) 83 *A.J.I.L.*
- Lauterpacht, International Law, quoted by I. Brownlie, *Principles of Public International Law* (Oxford: Clarendon Press, 1979).
- Lee, R.J., "Effects of Satellite Ownership Transfers on the Liability of the Launching States" (2000) 43 *IISL.*
- Leich, M. N., "U.S. Practice" (1982) 76 *A.J.I.L.*
- Longo, M., "Legal Aspects of Launching Space Objects from Non-Terrestrial Sites" (1999) 42 *IISL.*
- Loquin, E., "La gestion contractuelle des risques de l'exploitation commerciale de l'espace", in P. Kahn, ed., *L'Exploitation commerciale de l'espace: droit positif, droit prospectif*, (Dijon: Litec Credimi, 1992).
- Lyall, F., "UK Space Law", (1992) 35 *IISL.*
- Macdonald, R., "Recommissioning Law Reform 35" *Alberta L. Rev.*
- Madani, B., "New Report Links Syria to 1992 Bombing of Israeli Embassy in Argentina", Middle East Intelligence Bulletin, Vol. 2 No.

3, March 2000, http://www.meib.org/articles/0003_s1.htm accessed on November 1, 2000.

- Maiorsky, B., "Does the Convention on Registration of Objects Launched into Outer Space Require Revision? (1985) 28 *IISL.*
- Maniatis, D., "The Law Governing Liability for Damage Caused by Space Objects: From State Responsibility to Private Liability" (1997) XXII *Ann. Air & Sp. L.*
- Mann, F. A., *Further Studies in International Law* (Oxford: Clarendon Press, 1990).
- Marcoff, G., *Traité de Droit international public de l'espace* (Fribourg, 1973).
- Martin, P. M., "Legal Consequences of the Lack of French Space Legislation", (1991) 34 *IISL.*
- Masson Zwaan, T. L., "The Martin Marietta Case or How to Safeguard Private Commercial Space Activities" (1992) 35 *IISL.*
- Matte, N. M., *Aerospace Law From Scientific Exploration to Commercial Utilization* (Toronto: The Carswell Company Ltd., 1977).
- Matte, N. M., Space Activities and Emerging International Law (ed.) (Montreal: McGill University, 1984).
- Matte, N. M., *Space Policy and Programmes Today and Tomorrow* (Montreal: McGill University, 1980).
- Menard, B. P., "YUJI ISWASAWA, International Law, Human Rights, and Japanese Law: The Impact of International Law on Japanese Law Evidence of Compliance" (2000) 40 *Va. J. Int'l L.*
- Menter, M., "Legal Responsibility for Outer Space Activities", (1983) 26 *IISL.*
- Meredith, P. L. & G. S. Robinson, *Space Law: A Case Study for the Practitioner* (Dordrecht, Martinus Nijhoff, 1992).
- Meredith, P., "A Comparative Analysis of United States Domestic Licensing Regimes for Private Commercial Space Activities" (1989) 32 *IISL.*
- Meredith, P., "Implementing a Telecommunications Satellite Business Concept: Overview and Relative Timing of Legal Actions", (1990) 43 *IISL.*
- Meredith, P., "Risk Allocation Provisions in Commercial Launch Contracts", (1991) 34 *IISL.*
- Michalchuk, D., "Filling a Legal Vacuum: The Form and Content of Russia's Future State Immunity Law Suggestions for Legislative Reform" (2001) 32 *Law & Pol'y Int'l Bus.*

- Miklódy, M., "International Cooperation. A Legal Obligation in the Law of Outer Space?", (1983) 26 *IISL.*
- Mitchell, A.D., "Genocide, Human Rights Implementation and the Relationship between International and Domestic Law: Nulyarimma V Thompson" (2000) 24 *Melbourne U. L.R.*
- Mueller, B. M., "The Falkland Islands: Will the Real Owner Please Stand Up, (1983) *58 Notre Dame L. Rev.*
- Murphy, S. D. "*Prospective Liability Regimes for the Transboundary Movement of Hazardous Wastes*" (1994) 88 *A.J.I.L.*
- Mutti, A. H., Espacio Ultraterrestre. Política y Legislación del Espacio Exterior (Buenos Aires: Ediciones Particulares, 1997).
- Negro, S.C., Cooperación Espacial Comunitaria. Régimen Jurídico. Exploración y Explotación del Espacio (Buenos Aires: Ediciones Ciudad Argentina, 1997).
- Nesgos, P., "Recent Developments in Risk Allocation of Concern to the US Commercial Space Launch Industry and the Insurance Community", (1989) Assicurazioni Generali, Fifth International Conference on Space Insurance.
- Nesgos, P., "The Challenges Facing the Private Practitioner: Liability and Insurance Issues in Commercial Space Transportation" (1989) 4 *J.L. & TECH.*
- Nesgos, P., *National Law and Commercial Activities in Outer Space: Specific Aspects of United States Law Applicable to Space Industrialization with Particular Emphasis on the Commercialization of Launch Vehicle Transportation*, (DCL, Thesis, McGill University, 1983) [unpublished].
- Nolon, J. R., "Fusing Economic and Environmental Policy: The Need for Framework Laws in the United States and Argentina" (1996) 13 *Pace Envtl. L. Rev.*
- Nomura, T., "Japan's new long-term vision creating a space age in the new century", (1995) 11 *Sp. Pol.*
- Nordlund, Frédéric, Le régime juridiques des activités industrielles et commerciales conduites dans l'espace extra-atmosphérique: nouvelles orientations, (LL.M., Thesis, McGill University, 1990) [unpublished].
- Oberst, G., "Satellites and World Trade", (1999) Regulatory Update at 18.
- Ospina, S., "International Responsibility and State Liability in an Age of Globalization and Privatization. A Personal View of (established) Texts and (evolving) Contexts" *Fiftieth Anniversary Celebration Institute of Air and Space Law, McGill University*, April 19-21, 2002.
- Peyrefitte, L., *Droit de l'espace*, (Paris: Précis Dalloz, 1993)

- Piradov, A.S., (ed.), Droit Spatial International (Moscou: Progress Publishers, 1976).
- Ponzanelli, G., *La responsabilità civile* (Bologna: il Mulino, 1992).
- Postyshev, V., & I. Moiseyev, "Space Policy in Russia. Perspectives for Legal Development", (1992) 35 *IISL.*
- Quizi, H., "Certain Legal Aspects of Commercialization of Space Activities", (1990) XXII *Air & Space Law.*
- Ramey, R. A. "Armed Conflict on the Final Frontier: The Law of War in Space" (2000) 48 *A.F. L. Rev.*
- Raikes, C., Licensing Trends in Telecoms - The DTI Policies and Practices, *QDOS Telecommunications Law*, London, 9-10 February, 1998.
- J. L. Reed , "The Commercial Space Launch Market and Bilateral Trade Agreements in Space Launch Services" (1997) *13 Am. U. Int'l L. Rev.*
- Reynolds, G.H. & R.P. Merges, *Outer Space: Problems of Law and Policy*, (Westview Press, 1989).
- Rogoff, M. A., "Interpretation of International Agreements by Domestic Courts and the Politics of International Treaty Relations: Reflections on Some Recent Decisions of the United States Supreme Court" (1996) 11 *Am. U.J. Int'l L. & Pol'y.*
- Ruda, J. M., "The Role of the Argentine Congress in the Treaty Making Process" in S. A. Riesenfeld and F. M. Abbott (eds.), *Parliamentary Participation in the Making and Operation of Treaties: A Comparative Study* (Dordrecht: Martinus Nijhoff Publishers, 1994).
- Sagar, D., "Inmarsat since Privatization", Project 2001, Working Group on Telecommunication.
- Sagues, N.P., "An Introduction and Commentary to the Reform of the Argentine National Constitution" 28 *U. Miami Inter-Am. L. Rev.*
- Sanz Fernández de Córdoba, S., "Changing Basic Space Laws: Popularity, Pragmatism and Historical Lessons", (1993) 39 *IISL.*
- Sarkin, J., "The Effect of Constitutional Borrowings on the Drafting of South Africa's Bill of Rights and Interpretation of Human Rights Provisions" (1998) 1 *U. Pa. J. Const. L.*
- Sato, M., "The Japanese Legal Framework: Third Party Liability Resulting from NASDA Launch Activities", (1998) 41 *IISL.*
- Schacter, Who Owns the Universe? Space Law Symposium, Special Committee on Space and Astronautics, US Senate, 85th Congress, 2nd Session, 1957, 8-7 quoted by N. M. Matte, Space Activities and Emerging International Law (ed.) (Montreal: McGill University, 1984).

- Schadbach, K. "The Benefits of Comparative Law: A Continental European View", 16 *B.U. Int'l L.J.*
- Schmidt-Tedd, B. "Best Efforts Principle and Terms of Contract in Space Business" (1988) 31 *IISL.*
- Schwarzenberger, *International Law as Applied by International Courts and Tribunals* (London: Stevens & Sons Ltd., 1957).
- Shihata, I. F. I., "The Role of Law in Business Development", (1997) 20 *Fordham Int'l L.J.*
- Sidak, J. G., Foreign Investment in American Telecommunications, (Chicago: The University of Chicago Pres, 1997) at 135.
- Sinclair, I., "National Treaty Law and Practice: United Kingdom" in M. Leigh & M. R. Blakeslee, eds., *National Treaty Law and Practice* (Washington, DC: ASIL, 1995).
- Smith III, M. L., *International Regulation of Satellite Telecommunications after the Space WARC*, (D.C.L., Thesis, McGill University, 1989) [unpublished].
- Stein, E., "International Law in Internal Law: Toward Internationalization of Central-Eastern European Constitutions?" (1994) 88 *A.J.I.L.*
- Straubel, M. *United States' Regulation of Commercial Space Activity* (LL.M., Thesis, McGill University, 1989) [unpublished].
- Stuyt, M., *Legal Aspects of Commercial Activities of Private Enterprise in Outer Space*, (LL.M., Thesis, McGill University, 1985) [unpublished].
- Sucharitkul, S., "State Responsibility and International Liability under International Law" (1996) 18 *Loy. L.A. Int'l & Comp. L.J.*
- Tatsuzawa, K., "The Regulation of Commercial Space Activities by the Non-Governmental Entities in Space Law" (1988) 32 *IISL.*
- Templeman, L., "Treaty Making and the British Parliament" in S. A. Riesenfeld & F. M. Abbott (eds.), *Parliamentary Participation in the Making and Operation of Treaties: A Comparative Study* (Dordrecht: Martinus Nijhoff Publishers, 1994).
- Tomlinson, E. A. "Tort Liability in France for the Act of Things: A Study of Judicial Lawmaking", (1988) 48 *La. L. Rev.*
- Toope, S.J., "Re Reference by Governor in Council concerning Certain Questions relating to Secession of Quebec from Canada 161 D.L.R. (4th) 385. Supreme Court of Canada, August 20, 1998" (1999) 93 *A.J.I.L.*
- Trieschmann, J & R. Gustavson & G. Sandra, *Risk Management & Insurance*, 9th ed. (Cincinatti: South Western College Publishing, 1995).

- van Traa-Engelman, H. L., "Problems of State Responsibility in International Space Law", (1983) 26 *IISL.*
- Van Traa-Engelman, H.L., *Commercial Utilization of Outer Space* (Dordecht: Martinus Nijhoff Publishers, 1993).
- Vittadini Andrés, S.N., "First Amendment Influence in Argentine Republic Law and Jurisprudence" (1999) 4 *Comm. L. & Pol'y.*
- Vlasic, I., "A Survey of the Space Law Treaties and Principles Developed through the United Nations", (1995) 38 *IISL.*
- Volk, I. & A. Bauzá Araujo, "Space Transportation", in M. Folchi, ed., *40 Años de ALADA* (Buenos Aires: ALADA, 2000).
- von der Dunk, F., "Launching from Down Under: The New Australian Space Activities Act of 1998" (2000) (43) *IISL.*
- von der Dunk, F., "Public Space Law and Private Enterprise. The Fitness of International Space Law Instruments for Private Space Activities", (1998) *IISL.*
- von der Dunk, F.G., Private Enterprise and Public Interest in the European 'Spacescape' Towards Harmonized National Space Legislation for Private Space Activities in Europe (Leiden, IIASL, 1999).
- Warner, M. A. "Third Annual Latin American Competition and Trade Round Table: After Seattle: Is There a Future for Trade and Competition Policy Rule-Making?" (2000) 26 *Brooklyn J. Int'l L.* at 328.
- Wassenbergh, H., "International Space Law: A Turn of the Tide", (1997) XXII *Air & Space Law.*
- Wassenbergh, H., *Principles of Outer Space Law in Hindsight* (Dordrecht: M. Nijhoff Publishers, 1991).
- Watson, A., *Legal Transplants: an Approach to Comparative Law* (Athens, GA: University of Georgia Press, 1993).
- Wear, D. "INTELSAT: Evolving to Meet the Challenges of a New International Telecommunications Marketplace", (1995) 38 *IISL.*
- White, Stewart, S. Bate & T. Johnson, *Satellite Communications in Europe: Law and Regulation* (London: FT Law & Tax, 1996).
- Williams, C. D., "Space: The Cluttered Frontier", (1995) 60 J. Air L. & Com.
- Williams, Jr., C. A. & M. L. Smith, & P. C. Young, *Risk Management & Insurance*, 7th ed. (New York: McGraw-Hill Inc., 1995).
- Williams, M., *Report: Review of Space Law Treaties in View of Commercial Space Activities* (International Law Association: London, 2000).

- Williams, M., *Riesgo ambiental y su regulación* (Buenos Aires: Abeledo-Perrot, 1998).
- Williamson, M., "The UK Parliamentary Space Committee. The emergence of a lobby", (1992) 8 *Sp Pol.*
- Wirin, W.B., "Practical Implications of Launching State - Appropriate State Definitions", (1994) 37 *IISL.*
- Wirin, W. B., "Space Debris" (1989), 32 *IISL.*
- Yoshida, H., "The meaning of Japan's space commercialization efforts", (1992) 8 *Sp. Pol.*
- Zhukov, G.P. & Y. Kolossov, *International Space Law* (New York: Praeger, 1984).

List of acronyms

CNC: National Communications Commission
CNES: Centre National d'Études Spatiales
CNIE: National Commission of Space Research
CONAE: National Commission on Space Activities
COPUOS: Committee on the Peaceful Uses of Outer Space
CSG: Guiana Space Center
CSLA: Commercial Space Launch Act
ELA: Ensemble de Lancement Ariane
ELV: Expendable launch vehicle
ESA: European Space Agency
MPL: maximum probable loss
MTCR: Missile Technology Control Regime
NASA: National Aeronautics and Space Administration
PCS: Personal Communications Satellite
PSC: United Kingdom Parliamentary Space Committee
SIC: Space information cycle
UK: United Kingdom
UN: United Nations
US: United States
USA: United States of America

SPACE REGULATIONS LIBRARY

1. V. Kayser: *Launching Space Objects: Issues of Liability and Future Prospects.* 2001 ISBN 1-4020-0061-8
2. E. Sadeh (ed.): *Space Politics and Policy.* An Evolutionary Perspective. 2002 Hb ISBN 1-4020-0879-1; Pb ISBN 1-4020-0902-X
3. J. Hermida: *Legal Basis for a National Space Legislation.* 2004 ISBN 1-4020-1913-0

KLUWER ACADEMIC PUBLISHERS – DORDRECHT / BOSTON / LONDON

GPSR Compliance

The European Union's (EU) General Product Safety Regulation (GPSR) is a set of rules that requires consumer products to be safe and our obligations to ensure this.

If you have any concerns about our products, you can contact us on ProductSafety@springernature.com

In case Publisher is established outside the EU, the EU authorized representative is:

Springer Nature Customer Service Center GmbH
Europaplatz 3
69115 Heidelberg, Germany

Batch number: 10627899

Printed by Printforce, the Netherlands